The Thyristor Book
With 49 Projects

The Thyristor Book

With 49 Projects

Delton T. Horn

TAB BOOKS
Blue Ridge Summit, PA

FIRST EDITION
FIRST PRINTING

Copyright 1990 by TAB BOOKS
Printed in the United States of America

Library of Congress Cataloging in Publication Data

Horn, Delton T.
    The thyristor book—with 49 projects / by Delton T. Horn.
     p.   cm.
    ISBN 0-8306-8307-0   ISBN 0-8306-3307-3 (pbk.)
    1. Thyristors. I. Title.
    TK7871.99.T5H67 1990
    621.381'5287—dc20
                                                    90-35297
                                                    CIP

TAB BOOKS offers software for sale. For information and a catalog, please contact
TAB Software Department, Blue Ridge Summit, PA 17294-0850.

Questions regarding the content of this book should be addressed to:

    Reader Inquiry Branch
    TAB BOOKS
    Blue Ridge Summit, PA 17294-0214

Acquisitions Editor: Roland S. Phelps
Book Editor: Dan Early
Production: Katherine Brown

Contents

List of Projects

Introduction

NO ELECTRONICS HOBBYIST OR TECHNICIAN CAN HOPE TO SUCCESSFULLY
design or modify any circuit using components he doesn't understand.
Thyristors are not really uncommon, but they are poorly understood.
The literature available on these devices has been limited.

Thyristors are semiconductor components used primarily for
power control and switching applications. There are many different
types of thyristors. Most are comprised of four semiconductor layers.
The differences between various thyristors lies in the placement of the
leads. In a sense, thyristors could be considered specialized transistors.

This book is an in-depth examination of thyristors and their uses.
Both theory and practical applications are covered. Relatively common
devices such as SCRs, triacs, and UJTs are given the most space, but
rarer units, such as SCSs, SUSs, and LASCRs are also covered.

It is assumed that the reader has some prior experience in the field
of electronics. To get the most value out of this book, the reader should
understand basic semiconductor theory and the operating principles of
standard bipolar transistors. Many of the explanations in this volume
revolve around the functional differences between thyristors and "ordinary" bipolar transistors. The reader does not have to be an expert, or a
professional in the electronics field. Some review material is provided
where appropriate, but this book is not intended as a beginner's text.

The best way to fully understand how any electronic component
works is to experiment with it in practical "hands-on" circuits. For this
reason, almost fifty projects are included to illustrate the wide variety of
potential thyristor applications. Projects are a great way to learn about
electronic devices and circuitry. In addition, building projects can be a
lot of fun. Many of the projects featured in this volume will prove useful

for practical applications, and cost far less than comparable commercial units.

The thyristor projects in this book go well beyond simple light dimmers, although some light dimmer projects are included. You may be quite surprised at the versatility of thyristors.

❖1
Introduction to Thyristors

THYRISTORS ARE BOTH FAMILIAR AND MYSTERIOUS TO THE AVERAGE electronics hobbyist. Even the term *thyristor* itself is unknown to a great many hobbyists. It certainly sounds like a very imposing and intimidating word.

Sooner or later, most beginners do come across a thyristor or two, usually in kits or projects such as light dimmers or motor speed controllers. Such circuits are relatively simple and easy to build. Unfortunately, there is usually little or no explanation offered on the theory of just how the circuit works. These projects use transistorlike components known as *Silicon Controlled Rectifiers* (SCRs), and that's about all many of us know about them.

When you have finished reading this book, you'll know a lot more about the SCR and its various relatives that make up the thyristor family. And you'll be ready to use them in a wide variety of projects, including many that are fairly simple, and many more that are far more sophisticated than the simple beginner's lamp dimmer.

WHAT IS A THYRISTOR?

The term *thyristor* is used to identify an entire class of semiconductor devices. Thyristors are not entirely unlike the more familiar bipolar transistors used in so many electronics circuits. In fact, the name is a contraction of *thyratron transistor*. Sometimes, thyristors are called *reverse-blocking thyristors*, but this terminology is somewhat redundant, since all thyristors are, by definition, reverse-blocking.

Generally, thyristors are used to control the power fed to a load. In the early days of electronics, such applications were performed by large variable resistors and rheostats, and bulky, expensive thyratron tubes.

Thyratron tubes were seldom used unless they were absolutely necessary. It was usually more convenient and practical to avoid them.

Semiconductor thyristors, on the other hand, are compact and relatively inexpensive. They are also highly efficient and can often handle
very high currents. Some thyristor devices are rated for current handling capacities of as much as 4000 amperes. Since thyristors are so
small, they are often a very attractive alternative to relays or other elec-
tromechanical devices.

Ordinary semiconductor diodes have a single pn junction, and are
made up of two semiconductor layers, as illustrated in Fig. 1-1. Bipolar
transistors, are somewhat more complex devices. They have two pn
junctions and three semiconductor layers of alternating type, as illustrated in Fig. 1-2.

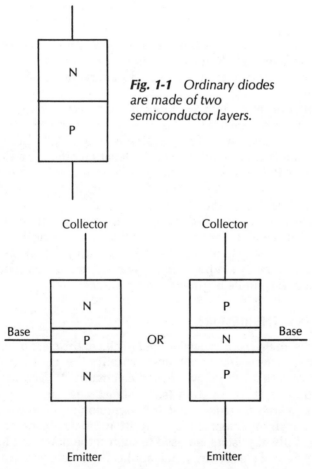

Fig. 1-1 *Ordinary diodes are made of two semiconductor layers.*

Fig. 1-2 *Bipolar transistors are made up of three semiconductor layers.*

Thyristors, as shown in Fig. 1-3, have four semiconductor layers of alternating type. That is, they consist of pnpn junctions. Notice that the placement of the device leads is not shown in this diagram. This is because there are several different thyristor devices, each with a different arrangement of leads. Most thyristors have three leads, although there are some exceptions.

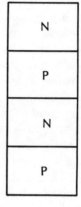

Fig. 1-3 *Thyristors are made up of four semiconductor layers.*

SEMICONDUCTOR JUNCTION DIODES

Before we go any further, it may be worthwhile to pause for a brief review of the basic semiconductor junction diode. If you feel that your background in semiconductor theory is strong enough, you may decide to skip over this section to the next one. Personally, I feel a little extra review will never hurt.

All materials conduct electricity to some degree. Materials such as silver or copper permit the current to flow through them very easily. Such materials, including most, but not all, metals, are known as *conductors*.

Other common materials, such as rubber and glass, are known as *insulators*. They greatly impede the current flow and make very poor conductors. An insulator acts as a very high electrical resistance.

A few less common materials, such as silicon and germanium, fall somewhere in between these two broad classifications. They don't make good conductors and they don't make good insulators. Such "in-between" materials are called *semiconductors*. Semiconductor materials are usually crystalline in makeup.

By itself, a semiconductor isn't really of much value in an electronic circuit. A piece of pure semiconductor material would function as nothing more than an overpriced medium-to-low value resistor.

For practical electronic semiconductor components, a controlled amount of specific impurities are added to the semiconductor crystals.

This process is known as *doping*. Every normal atom has an equal number of electrons and protons. Taking an electron away gives the atom a positive electrical charge, while adding an extra electron gives the atom a negative electrical charge.

If a slab of semiconductor material is doped with a substance that has more electrons than the semiconductor, the excess electrons will be free to roam about within the crystal. Because it has extra electrons, the doped crystal is called an *n-type semiconductor*.

It is also possible to dope a semiconductor crystal with a material that has fewer electrons than the semiconductor itself. The impurity atoms will tend to "steal" electrons from adjacent atoms in the crystal. Instead of moving electrons, we can describe the effect as moving holes, or spaces for electrons. Since a missing electron results in a positive electrical charge, a semiconductor crystal doped in this manner is called a *p-type semiconductor*.

Neither an n-type, nor a p-type semiconductor is particularly impressive or electronically useful on its own. But when we combine them, we can create some very powerful and versatile electronic devices for a wide variety of functions.

If we put a slab of n-type semiconductor against a slab of p-type semiconductor, as shown back in Fig. 1-1, we create what is known as a *pn junction*. Such junctions are at the heart of all of solid-state electronics.

Despite the extra electrons in an n-type semiconductor and the extra holes in a p-type semiconductor, the net electrical charge for both is zero. The doped crystal, as a whole, has an equal number of electrons and protons.

The excess holes and electrons come about because of the different atomic structures of the impurity elements added to the semiconductor. There are pockets of electrical charge scattered about within the crystal, but their overall effect is canceled. The net charge of the material as a whole is zero, or neutral.

The excess electrons and holes are called *carriers*. Both n and p semiconductor types have both type of carriers, although one or the other is much more plentiful in each type of doped semiconductor.

In an n-type semiconductor, electrons are the majority carriers and holes are the minority carriers; that is, there are more electrons than holes. This situation is reversed for a p-type semiconductor, where holes are the majority carriers and electrons are the minority carriers.

Figure 1-4 shows a typical pn junction. When no external voltage is applied, the carriers are randomly placed.

Now, let's suppose we've hooked up a voltage source across the pn junction, as shown in Fig. 1-5. The positive end of the voltage source is

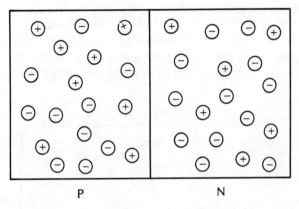

Fig. 1-4 *This is a typical pn junction.*

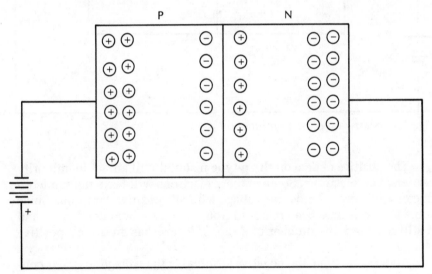

Fig. 1-5 *A reverse-biased pn junction*

applied to the n-type semiconductor, and the negative end is connected to the p-type semiconductor.

Under these circumstances, the holes in the p-type semiconductor will be drawn towards the end of the crystal with the most negative charge. Meanwhile, the excess electrons in the n-type semiconductor will be drawn off by the positive end of the voltage source.

Virtually no majority carriers will be found near the actual pn junction. As a result, virtually no electrons can jump across the junction from one type of semiconductor to the other. Almost no current can flow through the crystal. It exhibits a very high resistance. Under these conditions, the pn junction is said to be *reversed-biased*.

In Fig. 1-6, the polarity of the voltage source is reversed. Now, the positive connection is made to the p-type semiconductor and the negative voltage supply terminal is connected to the n-type semiconductor.

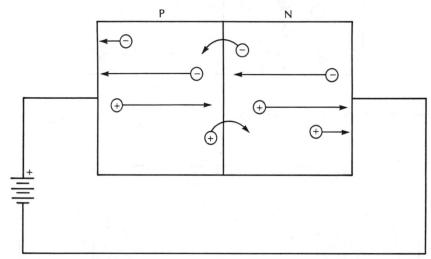

Fig. 1-6   *A forward-biased pn junction*

The positive charge on the p-type material will attract its minority carriers, i.e., electrons. Some of these electrons will leave the semiconductor and flow towards the voltage source's positive terminal. Since some electrons have been removed from the p-type semiconductor, and it still has the same number of protons, it now has an overall positive charge.

At the same time, the negative terminal of the voltage source is connected to the n-type semiconductor, repelling its majority carriers (electrons) towards the junction of the two semiconductor materials. Since there are now many electrons being pushed towards the junction, and a positive electrical charge pulling them from the other side, they are forced through the narrow junction area to neutralize the positively charged p-type side.

These junction-jumping electrons are now new minority carriers in the p-type semiconductor end of the crystal. Meanwhile, the voltage source is drawing away more electrons from the p-type material, so it retains its positive charge.

Clearly, current is flowing through the semiconductor junction under these conditions. The pn junction is now *forward-biased*.

You could, if you prefer, look at the whole procedure from the point of view of holes. The only difference is in the polarity, or direction of carrier movement. A hole, you'll remember, is just an empty space for an electron. If the electrons flow in one direction, the holes, by definition, flow in the opposite direction.

A simple pn junction permits current to flow in one direction, but not in the other. A component of this type is called a *diode*. The standard schematic symbol for a semiconductor diode is shown in Fig. 1-7.

The two ends of the diode are called the *anode* and the *cathode*. Current, the stream of electrons, flows from the cathode to the anode. Diodes are generally used for power rectification and polarity protection.

Fig. 1-7   *This schematic symbol is used to represent a semiconductor diode.*

The most important specification for a semiconductor diode is the *peak inverse voltage* (PIV). Sometimes this specification is referred to as *peak reverse voltage* (PRV). It means the same thing in either case. This is the maximum voltage which can be applied to a reverse-biased diode without the semiconductor junction breaking down.

With ordinary diodes, this value must never be exceeded, or the diode could be damaged or destroyed. In very simple terms, a thyristor is just a series of diodes (pn junctions), one right after another.

TYPES OF THYRISTORS

There are many different types of thyristors. Some of the most important varieties are listed below:

- SCR   (Silicon Controlled Rectifier)
- Diac
- Triac
- Quadrac

These devices are discussed in some detail in the first nine chapters of this book.

Less common, but still important thyristor devices include:

- ASR   Asymmetrical Silicon Rectifier
- CSCR   Complementary Silicon Controlled Rectifier
- SBS   Silicon Bilateral Switch

- SCS   Silicon Controlled Switch Shockley diode (four-layer diode)
- SUS   Silicon Unilateral Switch

Most of these less familiar thyristor devices will be explored in chapter 12.

Closely related to the thyristor family are unijunction devices:

- UJT   Unijunction Transistor
- PUT   Programmable Unijunction Transistor

These devices are covered in chapter 11.

Certainly, the most important type of thyristor is the silicon controlled rectifier, which will be discussed in the next chapter. Most other thyristors are variations on the basic SCR. Before we begin our examination of the SCR, we need to take a quick look at an unusual semiconductor device known as the four-layer, or Shockley diode.

THE FOUR-LAYER DIODE

The simplest of all thyristors is the four-layer diode. This device is also known as a Shockley diode. Like an ordinary diode, it has two leads, labeled *anode* and *cathode*. But instead of two semiconductor layers, as in a regular diode, this component is made up of four semiconductor layers. The internal structure of a four-layer diode is illustrated in Fig. 1-8. The most commonly used schematic symbol for a four-layer diode is shown in Fig. 1-9.

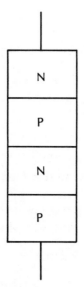

Fig. 1-8 *The four-layer diode is a two-lead thyristor.*

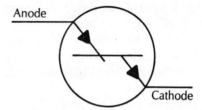

Fig. 1-9 *This is the schematic symbol for a four-layer diode.*

Incidentally, the term *Shockley diode* sounds complex and mysterious, but actually it's quite simple. The inventor of the four-layer diode happened to be named Shockley, so his name was given to the component he created.

When reverse-biased, the four-layer diode is not unlike an ordinary two-layer diode. The resistance is high, permitting very little current to flow through the diode. But when forward-biased, the four-layer diode behaves in a very unusual manner. As the applied voltage is increased, the current flow will remain low until a specific critical point is reached. At this point, the current will suddenly rise quite steeply. The current will continue to rise, even if the forward voltage is reduced almost to zero.

The four-layer diode exhibits negative resistance and switching properties, comparable to a neon glow lamp. Each of these important properties are described below.

Negative Resistance

People working in electronics are quite familiar with positive resistance. This is a resistance that functions according to Ohm's Law:

$$E = IR$$
$$I = E/R$$
$$R = E/I$$

The voltage increases with increases in the current flowing through the resistive element.

A negative-resistance device, like the four-layer diode, does not obey Ohm's Law. In a negative-resistance device, increasing the current causes the voltage to decrease rather than increase.

If the applied voltage is within a specific range, determined by the specific characteristics of the individual device, oscillation will result.

Switching

When reverse-biased, a four-layer diode is held in its OFF state. It will exhibit a very high electrical resistance, like any reverse-biased diode. If the four-layer diode is forward-biased with a relatively low voltage, it will still stay OFF.

As the forward voltage is increased, a certain critical point will be reached. This critical voltage, which will be identified on the manufacturer's specification sheet for the specific device, may be known by any of several names:

- trigger voltage
- switching voltage
- threshold voltage
- avalanche voltage
- firing voltage

These various terms are virtually interchangeable.

When the applied voltage reaches the trigger voltage level, the four-layer diode suddenly switches over to its ON state. The electrical resistance now exhibited by the device drops to a very low level—typically just a few ohms.

A brief trigger pulse (a forward-biasing voltage pulse) is used to switch a four-layer diode ON. To switch the device back OFF again, the applied voltage must be removed altogether, or reversed. That is, the four-layer diode can be turned OFF by reverse-biasing it.

The switching time of this type of device is incredibly fast. In some four-layer diodes the switching time is on the order of a few nanoseconds. In case you are not familiar with the term, a nanosecond is a billionth of a second. Generally the switching time to turn the device ON is somewhat shorter than the time required to turn it OFF.

PROJECT 1: FOUR-LAYER DIODE OSCILLATOR

A four-layer diode can easily be forced into oscillation. Figure 1-10 shows a very simple oscillator circuit built around this device. An oscillator circuit couldn't be much simpler than this. Other than the four-layer diode itself and the load resistance (output device), the circuit calls for just two components: a resistor and a capacitor. This type of oscillator circuit is known as a relaxation oscillator.

When power is applied to the circuit, capacitor C1 is charged through R1. At some point, the charge on the capacitor will exceed the trigger voltage of the four-layer diode. The diode turns ON and permits current to flow through load R1.

In addition, while the four-layer diode is in its ON state, the capacitor is shorted across the diode and the load resistance. Of course, the capacitor starts to discharge. When the current from the capacitor falls

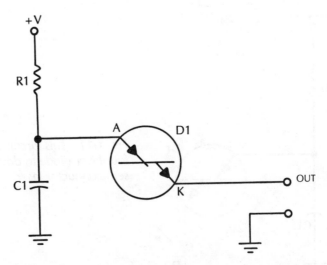

Fig. 1-10 *A four-layer diode can easily be forced into oscillation.*

below a specific minimum level, the cut-off voltage of the four-layer diode, the diode will snap back to its OFF state.

The situation is now the same as what we started with. Capacitor C1 starts to charge through resistor R1 again, and the entire process repeats.

The values of resistor R1 and capacitor C1 determine how fast the capacitor charges up to the trigger voltage, so these component values can be changed to control the output frequency of the oscillator. Experiment with different component values.

PROJECT 2: FOUR-LAYER DIODE LASER DRIVER

Another fairly simple circuit using a four-layer diode is shown in Fig. 1-11. This circuit is a simple driver network for a laser diode. Most laser diodes are fired by a string of pulses, rather than continuously. This is done to limit power consumption and heat buildup. This circuit provides the necessary pulses to drive a typical semiconductor laser diode.

Basically, this is a variation on the oscillator circuit shown in Fig. 1-10. For this circuit the capacitor should have a value of about 0.01 μF. 22K is a good value for the resistor. These parts values aren't too critical, but they should be in this general range for proper operation with a laser diode.

Using these component values, the laser will be fired by pulses with a frequency of about 1 kHz (1000 Hz). The pulse width is approxi-

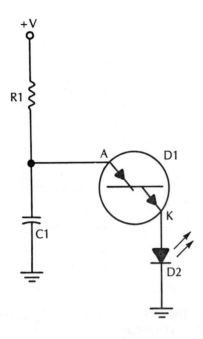

Fig. 1-11 *This circuit uses a four-layer diode to drive a semiconductor laser.*

mately 100 nanoseconds (0.000001 second). This circuit's output is close to 10 amps on pulse peaks. A laser diode needs a high driving current to function.

If you decide to experiment with laser diodes, use all precautions. Be careful where you aim the beam. Keep the project away from children. Never look directly into a laser beam.

If you don't know what you're doing, it's not advisable to be careless with laser diodes. If you happen to be interested in this area, a number of books on lasers are available: *Lasers—the Light Fantastic—2nd Edition* (TAB Book #2905, with Clayton L. Hallmark), *Laser Experimenter's Handbook—2nd Edition* (TAB Book #3115) and *Understanding Lasers,* by Stan Gibilisco (TAB Book #3175).

❖2
SCRs

WHILE THE FOUR-LAYER DIODE, DISCUSSED IN CHAPTER 1, HAS TWO LEADS, most common thyristor devices have three leads, like a transistor.

If we add a third lead to a four-layer diode, as shown in Fig. 2-1, we have a Silicon Controlled Rectifier, or SCR. The SCR is undoubtedly the most popular and common type of thyristor.

This component is pretty easy to understand, if we look at the individual words in its name. "Silicon," of course, is the semiconductor material the SCR's active portion is made of. "Rectifier" is also perfectly straightforward. It indicates that the device is a diode-like component, capable of rectification. It is actually a modified four-layer diode. The important term here is the one in the middle—"Controlled."

The rectification of an SCR is externally controllable via the added third terminal, called the "gate." The other two terminals carry over the names "anode" and "cathode" from regular two-terminal diodes. You could consider an SCR to be an electrically switchable diode.

The most widely used standard schematic symbols for an SCR are illustrated in Fig. 2-2. The use of the surrounding circle is entirely optional. Some technicians think the circle makes the symbol more clearly visible, while others feel it doesn't add any information to the diagram, and so is unnecessary. This is purely a matter of personal preference. Both of these symbols are in widespread use today.

The only real restriction in the choice of which of these schematic symbols to use, is that all of the symbols in any given schematic diagram should be consistent. Don't mix symbol types, or you will be begging for confusion.

The SCR was originally designed as a solid-state counterpart to the thyratron tube, mentioned in chapter 1. Like the thyratron tube, the

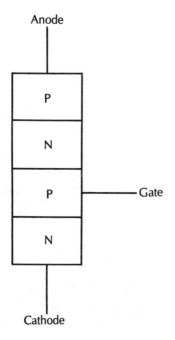

Fig. 2-1 *Adding a third (gate) lead to a four-layer diode creates an SCR.*

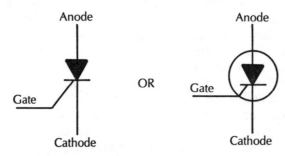

Fig. 2-2 *This is the schematic symbol for an SCR.*

SCR, and most modern thyristor components, is generally used in ac or dc power control applications.

If a voltage is applied between the anode and the cathode of an SCR, as shown in Fig. 2-3, nothing will happen as long as there is no signal on the gate lead. In Fig. 2-3, the gate lead is shown grounded to emphasize that there is no signal being applied to this terminal. This is not a practical circuit. It is included here solely to demonstrate the operating concepts behind the SCR. The circuit of Fig. 2-3 is not functional, and does absolutely nothing. The SCR, in this circuit acts like an open circuit.

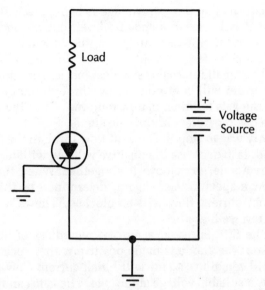

Fig. 2-3 *Without a signal on the gate lead, no current can flow between the anode and the cathode.*

But what happens if we apply a voltage signal to the gate of the SCR, as illustrated in Fig. 2-4? If the gate voltage is very low, there will be no change. The SCR continues to block current flow from its cathode to its anode.

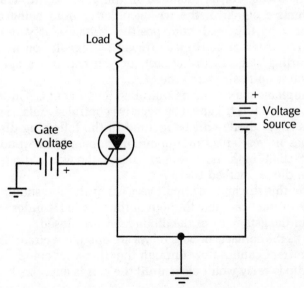

Fig. 2-4 *An SCR can be turned on by applying a signal to its gate terminal.*

Let's assume that the gate voltage is being gradually increased. At some point it will exceed a specific level, determined by the internal construction of the SCR. At this point, the SCR will be triggered. The "rectifier" is activated. Current can now flow from cathode to anode against only a small internal resistance, just as with an ordinary diode.

This current will continue to flow through the device, even if the voltage on the gate terminal is now removed. Once the SCR is switched ON, it stays ON, regardless of the gate signal.

The only way to stop the current flow through the SCR, once it has been started, is to decrease the positive voltage, with respect to the cathode, on the anode, or remove it altogether. When the anode voltage drops below a specific level, again, determined by the SCR's internal construction, current flow will be blocked. The SCR turns itself OFF when it is reverse-biased.

Once the SCR is OFF, it stays OFF, regardless of the voltage at the anode. Even if the anode is made positive with respect to the cathode, the SCR will remain OFF and will block current flow until it is again triggered by a suitable voltage at the gate. The gate can turn the SCR ON, but it cannot turn it OFF. Essentially, an SCR is an electrically switchable diode. This component is sometimes referred to as a "programmable diode."

Conceptually, you can consider an SCR as a latching relay in series with a diode, as illustrated in the equivalent circuit of Fig. 2-5. Notice that there are actually two relays in this circuit. They are labeled the "gate relay" and the "diode relay."

An input diode, D2, is included in the gate circuit. This diode merely prevents a signal of the wrong polarity from being applied. SCRs can only be triggered from positive voltages. If your specific application requires a negative gate triggering signal, you must use a polarity inverting stage in the circuit, or you can use a specialized device, which is quite similar to the SCR.

The Complementary Silicon Controlled Rectifier or CSCR, operates like an SCR, except it is triggered by negative control signals. (See chapter 12.) This input diode will be ignored in the following discussion because it has no real effect on the circuit's operation beyond simple polarity protection. In the next few paragraphs, the word "diode" refers to the output diode, marked D1 in Fig. 2-5.

Look over this illustrative circuit very carefully. You should be able to see that no current can flow through output diode D1 unless the contacts of either the gate relay or the diode relay are closed.

As long as the contacts of both relays are open, the circuit is deactivated, and current cannot flow through the diode. Of course, the contacts of the diode relay won't close until the coil is energized, and this

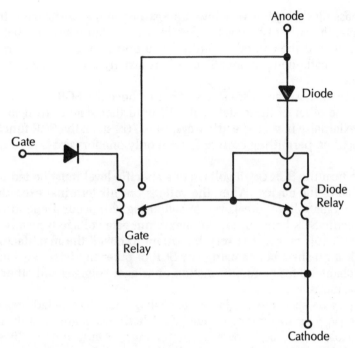

Fig. 2-5 *An SCR functions as a latching relay in series with a diode.*

can't happen unless current is flowing through the diode. So the only way to activate the diode relay is to activate the gate relay.

When the gate relay is activated by a sufficient voltage at the gate input, it will cause the contacts of the diode relay to close too. The diode relay's contacts are held closed by the current flowing through its coil, which is the same as the current flowing through the diode. Once the diode relay is activated, the state of the gate relay is irrelevant.

If the input gate signal is removed, the gate relay will be deactivated and its contacts will reopen. But this doesn't matter. The current flowing through the diode will hold the contacts of the diode relay closed.

If current through the diode stops flowing or changes direction, reverse-biasing the diode, there will be no more current flowing through the diode relay's coil either. The diode relay will de-energize, and its contacts will open. This simple dual relay circuit should give you a rough but useful idea of the basic functioning of an SCR.

ON AND OFF STATES

The SCR has two stable states. These states are usually called "ON" and "OFF." In the ON state, current can flow from the anode to the cath-

ode, passing through a very low, almost negligible, resistance. In the
OFF state, the current flow is blocked by a very high resistance between
the anode and the cathode. Neither state normally permits current to
flow from cathode to anode. The SCR is strictly a dc device, like any
rectifier.

Both of these states are stable. That is, once the SCR is placed into
one or the other of these states, it will hold that state until it is acted
upon externally in very specific ways. In the ON state, the SCR functions
as a rectifier, permitting current flow in only one forward-biased direction.

To turn the SCR ON, a voltage of a specific level must be fed to the
gate lead of the device. When the voltage on this terminal exceeds the
unit's trigger voltage, breakover, or switching will occur. Depending on
the specific SCR being used, the maximum gate voltage typically is in
the 2 to 5 volt range. It is very important to check the manufacturer's
specification sheet before using any SCR or other thyristor. See chapter
6 and chapter 9 for more information on trigger voltages and other SCR
specifications.

On some spec sheets, the trigger voltage may be labeled breakover
voltage (V_{bo}) or switching voltage (V_s). These terms are entirely interchangeable. Don't be thrown by the change in terminology. In some
areas of electronics, unfortunately there isn't much standardization.

Different technicians and manufacturers prefer different labels.
Since the choice is ultimately pretty arbitrary, it would seem to make
sense for everyone to get together and agree on one fixed term for each
parameter, but this is unlikely to happen in the near future. Anyone
working in electronics, either professionally or as a hobby, needs to be
aware of such inconsistencies in terminology.

In a practical circuit, the SCR should be rated for a maximum voltage higher than the peak voltage it will ever be exposed to in the circuit.
The circuit designer should always leave plenty of extra "elbow room."
Otherwise, an excessive voltage fed through the SCR could damage or
destroy it. This could also be harmful to any load circuit being driven by
the SCR.

As a general rule, the SCR should be rated for *at least* 10 to 25 percent higher than the absolute maximum anticipated voltage in the circuit. More is better. You can't use too high a voltage rating for a
thyristor. The only trade-offs are cost and the physical size of the component itself.

There is also a very real risk of hazards to the circuit operator. Refer
to chapter 7 for details on safety considerations for using SCRs and
other thyristors.

The maximum acceptable voltage that can safely be fed through a
given SCR is generally labeled "V_{drm}" on manufacturer's specification

sheets. This rating is more-or-less equivalent to the peak reverse voltage rating for a standard rectifier diode. As a rule, the V_{drm} rating will be at least 100 volts greater than the V_{bo}, or V_s rating.

In selecting an SCR for a specific application, leave plenty of "headroom" in the V_{drm} rating. For example, if you are working with ac line current (120 volts ac), it would not be unreasonable to use an SCR rated for at least 600 volts.

Taking a somewhat different conceptual approach, Fig. 2-6 shows another rough equivalent circuit for an SCR. This circuit makes more of a comparison to a standard semiconductor operation.

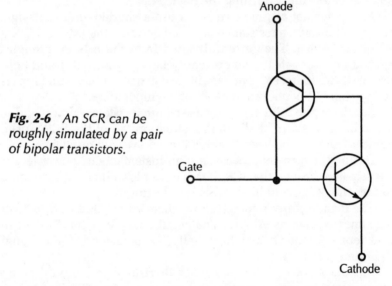

Fig. 2-6   *An SCR can be roughly simulated by a pair of bipolar transistors.*

Here we have two bipolar transistors interconnected so that if a current is injected into any leg of the combination, it will be amplified, or regenerated in the opposite leg. The gain of the transistors in this circuit is assumed to be relatively high for this to work. The sum of the common-base current gains (alpha, or α) for the two transistors must exceed unity (be greater than 1).

Junction leakage-current is relatively low, and the current gain of the circuit is designed to be low at the leakage-current level, so the overall device, which is effectively a pnpn transistor, remains OFF unless an external signal is provided.

When a sufficient trigger current is applied to the gate of Fig. 2-6, the loop gain is raised to unity. This allows regeneration to occur. The combination pnpn transistor is switched ON. The principal current through this circuit is limited primarily by the load impedance, or by resistances in the external circuitry.

If the trigger current is now removed, nothing happens. The double-transistor remains ON, provided that there is enough current to ensure latching by keeping the gain above unity. To turn the circuit OFF, the gain must be dropped below the unity level.

For the rough demonstration circuit shown in Fig. 2-6, it would appear that shorting the gate to the cathode should do the trick. However, this won't work in a practical SCR circuit. This is just an inaccuracy in the equivalent circuit. It is not fully equivalent to the real thing. In an actual SCR circuit, a gate – cathode short won't divert much current. In practice, the principal current, the current flowing between the anode and the cathode, must fall below a specific level before the gain can drop below unity, turning the device OFF.

Many, but not all, modern SCRs use a shorted-emitter design. In this type of design, a resistance is added between the gate and the cathode of the device. The current diverted from the n base through this resistance causes gate trigger current I_g, latching current I_l, and holding current I_h to increase. The shunt resistance's chief function is to improve the SCR's performance at high temperatures.

Sensitive gate SCRs typically use a high resistance shunt, and some use no internal shunt at all. In this situation, all of the SCR's primary characteristics can be heavily altered by an external resistance.

During the turn-ON process, a high instantaneous power level will be dissipated through the SCR. If this power level increases too much or too fast, the SCR could be damaged or destroyed.

The manufacturer's specification sheet will include a dI/dT rating, sometimes written as $\Delta I/\Delta T$. The small triangle or the "d" in these terms represent the Greek letter "delta," which stands for a changing, rather than a static value.

In working with SCRs and other thyristors, you should be aware that junction-leakage currents and current gain will increase with increases in temperature. The main effect of this is that the device can be triggered by a lower gate current. The gate itself may be regarded as a diode (refer to Fig. 2-5) and will display a lessening voltage drop with increases in temperature.

THE SCR IN ACTION

A number of practical projects using SCRs and other thyristors will be presented in chapter 10. For the remainder of this chapter we will look at a few very simple SCR circuits to demonstrate just how this component works.

Breadboarding

It is recommended that you breadboard and experiment with the circuits presented in the remainder of this chapter. Each of these circuits illustrate various aspects of SCR operation, and you can learn a great deal about thyristors by working with these circuits.

There would be no point in soldering together permanent versions of any of these projects. They are intended only to demonstrate principles, not to serve practical applications. Put these circuits together on a solderless breadboard so you can easily experiment with different component values.

A solderless breadboard is a special socket with openings for a number of component leads. Many different types of components may be mounted on such a socket, including ICs, transistors, thyristors, resistors, diodes, and capacitors. Components can easily be removed and replaced in the circuit. Tiny internal spring clips hold the leads in place.

Within the solderless socket, the holes are electrically interconnected in a specific pattern. Usually, all component leads in a single row are shorted together.

Never make any changes in a breadboarded circuit with power applied. Be careful that no exposed component leads touch one another, causing unintended shorts.

PROJECT 3: SIMPLE DEMO CIRCUIT

A simple SCR demonstration circuit is shown in Fig. 2-7. In this circuit we have an external switch in series with a current-limiting resistor which is used to open and close the gate current path. The series resistor in the gate circuit ensures that the gate current will drop to a safe level once the SCR has been triggered.

A 1N4004 diode is also installed in series with the gate in this circuit. This diode blocks the negative half-cycles of the incoming ac waveform. A large negative voltage on the gate lead could possibly damage the gate-cathode junction of the SCR.

In our analysis of the action of this circuit, we will assume that the SCR is in its OFF state when we begin. This is only a matter of convenience in our discussion. If we started out from an ON state initially, only the sequence of the events discussed below would be changed.

As long as the gate switch is open, nothing happens. The SCR remains OFF. Closing this switch feeds a signal to the gate and the SCR is turned ON during the positive half-cycle of the ac line voltage driving

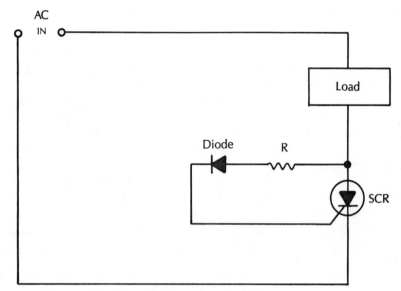

Fig. 2-7 *This is a simple SCR demonstration circuit.*

the circuit. The SCR will remain in the ON state as long as the forward current through it exceeds the holding current (I_h) value.

Once the SCR has been turned ON, the current flowing through it will be limited entirely by the external load circuit, or device being controlled by the SCR.

Since the signal in this circuit is an ac waveform, specifically, a sine wave, the current across the SCR will rise to a positive peak, then start to drop off, eventually passing through zero and going on to a negative half-cycle, which is a mirror image of the positive half-cycle. At some point in the cycle near the zero crossing point the current will fall below the I_h level, and the SCR will be turned OFF. It remains OFF throughout the negative half-cycle.

Assuming the gate switch is left closed, the SCR's gate will be retriggered after the next positive half-cycle begins, and the entire process described above repeats.

The SCR must be switched ON, or retriggered, once per cycle while the anode is positive with respect to the cathode. The SCR will turn itself OFF once per cycle on the negative half-cycles which make the cathode more positive than the anode, reverse-biasing the SCR.

Typical input and output signals for this circuit are illustrated in Fig. 2-8. Compare this diagram with the signals displayed on an oscilloscope, if at all possible.

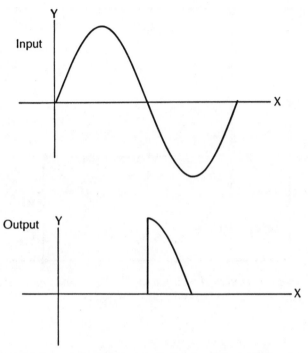

Fig. 2-8 *Typical input and output signals for the circuit of Fig. 2-7.*

PROJECT 4: SCR DEMONSTRATION CIRCUIT WITH ADJUSTABLE TRIGGER

Figure 2-9 shows the same basic circuit with a simple, but important modification. Here the series resistance in the gate circuit is made variable. By changing this resistance value, we can control just when during each positive half-cycle the SCR will be triggered. This, in turn, controls the average power applied to the load at the output of the circuit. Some typical input and output signals are illustrated in Fig. 2-11.

The adjustability of this demonstration circuit makes it closer to SCR circuits designed for practical applications. Most SCR circuits are variations on this basic circuit.

PROJECT 5: SIMPLE AC POWER CONTROLLER CIRCUIT

The circuit shown in Fig. 2-12 is a slightly more sophisticated version of the simple demo circuit presented earlier.

In this circuit the control voltage to the gate is applied by firing a

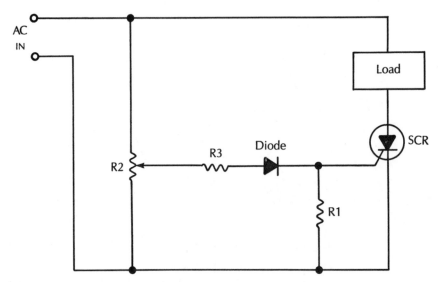

Fig. 2-9   *This variation on the circuit of Fig. 2-7 permits manual control over when, in each cycle, the SCR will be switched on.*

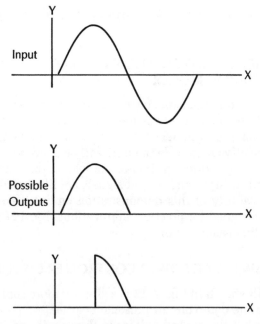

Fig. 2-10   *Some typical input and output signals for the circuit of Fig. 2-9.*

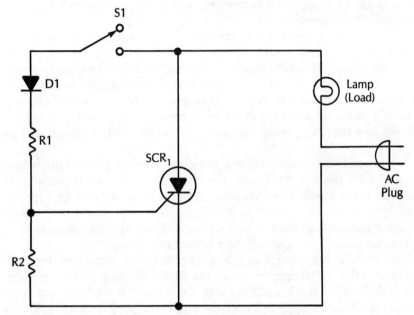

Fig. 2-11   *This is a simple ac power controller circuit.*

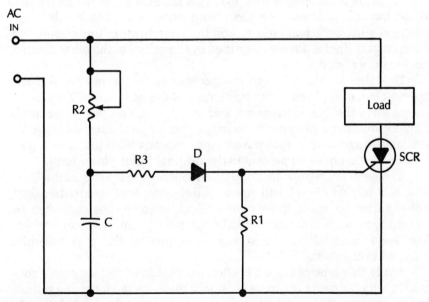

Fig. 2-12   *A phase-shift network can be used to extend the possible turn-on range within each cycle.*

neon lamp. This approach is frequently used in practical SCR circuits, especially in older designs.

The ac output power fed to the load is controlled by varying the firing time of the SCR. When, during the positive half-cycle the SCR is triggered, determines how much of the waveform and the average power gets through to the output load. Since there is an incomplete waveform at the output, the rms power is effectively reduced. At least half, and possibly more, of each input cycle is rejected. The output power is always less than 50 percent of the input power with a circuit of this type.

If the SCR is triggered right at the zero-crossing point at the beginning of each positive half-cycle, the circuit will put out its maximum possible output power. In this case, the output power will be approximately one-half of the input power. The effect of the circuit in this case is about the same as simple half-wave rectification. The negative half-cycles are simply stripped off from the output waveform.

If the SCR is triggered at any other point during the positive half-cycle after the initial zero crossing, the SCR will conduct for a proportionately smaller section of the cycle. Obviously, it will conduct from the point when it is triggered until the end of the positive half-cycle (the next zero crossing). Since more of the waveform is stripped off, the output power is correspondingly reduced.

A simple neon lamp is used to trigger the SCR's gate in this circuit. When the voltage across the neon lamp exceeds a specific value, the neon gas within the lamp reacts, and the lamp fires. In this circuit, the neon lamp is fired at a rate determined by the values of the other components in the circuit.

By using a manually variable potentiometer for resistor R1, the operator can easily control the firing time of the neon lamp. This determines when the SCR is triggered, and therefore the output power level. The potentiometer can adjust the output power level smoothly over its available range. The output power range extends from just over 0 percent to a little under 50 percent of the original input power level.

The neon lamp provides solid, unambiguous triggering of the SCR. The SCR in this circuit will not be falsely triggered by any transient noise pulses on the ac power lines. Such noise pulses can often be caused by motor brushes and by some other electro-mechanical devices. The neon lamp also reduces any variations in the gate potential required to trigger the SCR.

While this type of circuit is often referred to as an "ac power controller," the output is not truly ac. It is actually in the form of a pulsating dc voltage. This may be a problem for certain specialized loads. In such cases it will be necessary to use a triac—see chapter 3.

DC TRIGGERING

In some circuits, an SCR does not necessarily have to be triggered by part of the ac waveform. It can also be triggered by a pulse or a dc voltage. This offers the advantage that the turn-on point does not have to be between the 0° and 90° points of the waveform. It can occur at any point. On the other hand, timing the dc gate pulse to a specific point in an ac waveform can be somewhat tricky.

PROJECT 6: PHASE-SHIFT TRIGGERED SCR

One technique is to use a phase shift network, as shown in the circuit of Fig. 2-13. The phase shift network is made up of a resistor and a capacitor. Using this circuit, the turn-on point can be set for anywhere form 0° to 180° of the input cycle.

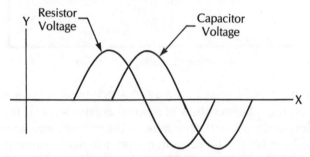

Fig. 2-13 *In the phase-shift circuit of Fig. 2-12, the voltage across the capacitor will be 90 degrees behind the voltage across the resistor.*

Capacitor C1 is placed between the SCR's gate and cathode. Figure 2-14 compares the phase relationship of the voltage across C1 with the input voltage fed to the R2/C1 network. The voltage across the capacitor is called V_{gc} here.

If you remember your basic electronics theory, you know that the current in a capacitor leads the voltage by 90°. Since a resistor doesn't care if it is carrying ac or dc, the voltage across this component is always in phase with the current flowing through it.

By combining a capacitor and a resistor in series and feeding them an ac waveform, the voltage across the capacitor will be 90° behind the voltage across the resistor.

The sum of the voltages dropped by the resistor and the capacitor must always equal the voltage fed to these two components (V_{ac}). This is due to one of the most fundamental laws in all of electronics. But things are complicated by the fact that we can't use simple addition under these circumstances. We must take the phase angles into account.

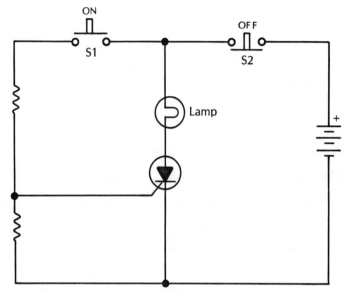

Fig. 2-14 *This circuit is a simple dc power controller.*

The total effective phase shift of the RC network as a whole is dependent on the relative values of the two components. If the resistor has a very large value, it will drop most of the voltage, and there will be very little voltage across the capacitor. Under these circumstances, the overall phase shift will be about 90°.

By adjusting the value of R2, we can control the voltage across capacitor C1. The voltage across the capacitor, V_{gc}, determines the firing time of the SCR.

DC POWER CONTROLLERS

SCRs seem to be most commonly used in ac circuits, but there is no particular reason why they can't be used to control dc circuits as well.

PROJECT 7: FIRST DC POWER CONTROLLER

A simple dc power controller circuit is shown in Fig. 2-14. Switch S1, an SPST Normally Open pushbutton type switch, is momentarily closed to trigger the SCR and turn ON the output device, the lamp in this case. When this switch is closed it feeds a voltage to the SCR's gate through resistor R1.

The cathode is tied to ground potential through resistor R2 for stability. This won't always be necessary in all applications, but it is usually advisable as cheap stability insurance. In a practical circuit, there

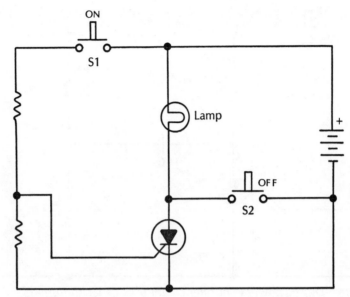

Fig. 2-15 *This dc power controller circuit uses a somewhat different approach than in Fig. 2-14.*

is rarely any strong reason not to ground the cathode through a resistor in this manner.

Of course, once the SCR has switched on, the gate voltage may be removed and the SCR will remain in its ON state. This is why a momentary action switch, S1, is used to trigger the circuit. A brief input pulse is all that is required. Once the SCR has been triggered, releasing (opening) switch S1, or pushing (closing) it again has no effect.

Once an SCR has been turned ON, the only way to turn it OFF is to drop the current flowing through it below the holding current (I_h). Switch S2 is included in the circuit for this purpose. This switch is an SPST Normally Closed pushbutton type. Momentarily opening this switch turns OFF the SCR and its load, the lamp. When this switch is opened, the circuit's power source is simply disconnected from the SCR, effectively reducing the current flow to zero. Depressing this switch while the SCR is in its OFF state has no effect.

PROJECT 8: ALTERNATIVE DC POWER CONTROLLER

A slightly different dc power controller circuit appears as Fig. 2-15. This circuit is quite similar in many ways to the one shown in Fig. 2-14. Once again, pushbutton switch S1 is momentarily closed to turn ON the SCR and its load, the lamp.

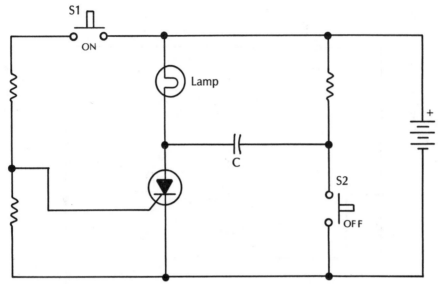

Fig. 2-16   *This dc power controller circuit uses capacitor turn-off.*

The method used here to turn OFF the SCR is a little different than in the earlier circuit. This time an SPST Normally Open pushbutton type switch is used to turn the circuit OFF. Momentarily closing this switch shorts the SCR's anode to its cathode (the cathode is grounded).

In other words, closing the "OFF" switch, S2, effectively grounds the anode. This has exactly the same effect as cutting off the supply voltage—the current flowing through the SCR is immediately reduced to zero, and it is switched to its OFF state.

PROJECT 9: CAPACITOR TURN-OFF DC POWER CONTROLLER

Still another approach is illustrated in Fig. 2-16. As in Figs. 2-14 and 2-15, switch S1, a momentary-action type, is used to turn the SCR and its load (lamp) ON by feeding a voltage to the gate when the switch is briefly depressed. Once again, the difference in this circuit lies in the method for turning the SCR and its load OFF.

The method of SCR turn-OFF used in the circuit of Fig. 2-16 is commonly called "capacitor turn-OFF." Notice that a Normally Open momentary-action pushbutton switch, S2, is used to control the OFF function in this circuit.

Once the SCR has been turned ON, capacitor C1 will be charged through resistor R3. It will soon reach a charge value just under the supply voltage. Then it will sit there until OFF switch S2 is momentarily closed. When this switch is closed, the positive end of capacitor C1 is

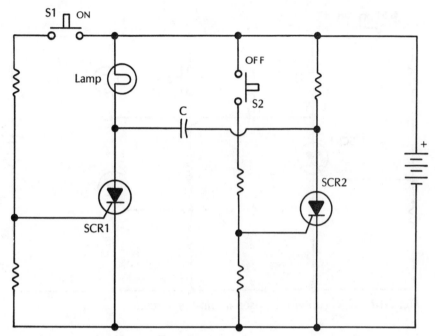

Fig. 2-17 A "slave" SCR can be used to turn OFF the main SCR.

shorted to ground, causing it to quickly discharge. This drives the anode of the SCR negative for a moment.

The SCR is reverse-biased by this negative voltage on its anode, so it reverts to its OFF state. The capacitor discharges very quickly, so the anode is not held negative for very long. But the anode only has to be held negative for a few microseconds to ensure turn-OFF, so the job is efficiently done by this method.

For this circuit, the capacitor must be a relatively large value (above 1 μF) nonpolarized type. It does not have to be too large. Anything over 10 μF probably would be overkill in almost all practical applications. Generally, a good range of values for this component would be 2 μF to 5 μF.

PROJECT 10: SLAVE OFF DC POWER CONTROLLER

Figure 2-17 shows a circuit which takes the capacitor turn-OFF idea one step further. Here we have a second "slave" SCR to turn OFF the main (load-controlling) SCR. The "main" SCR is labeled Q1 and the "slave" SCR is labeled Q2.

SCR Q1 is turned ON in the same way as all of the preceding circuits in this section. To turn SCR Q1 OFF, the slave SCR must be turned ON by

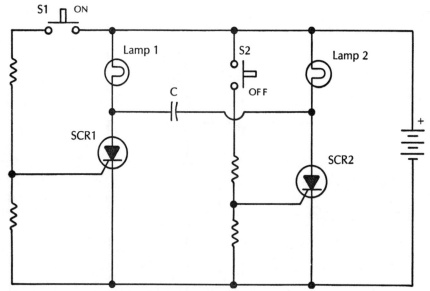

Fig. 2-18 *SCRs can also be used in flip-flop circuits.*

momentarily closing switch S2. SCR Q2 effectively takes the place of switch S2 in Fig. 2-16. When SCR Q2 is ON, capacitor C1 can discharge through it, causing the anode of SCR Q1 to go negative, turning Q1 OFF.

How do we turn OFF SCR Q2 then? Just release (open) switch S2. The anode current provided by resistor R3 is below the SCR's holding current, I_h, so the SCR is turned OFF as soon as switch S2 is opened.

PROJECT 11: FLIP-FLOP

Another way to use two SCRs in a single circuit is shown in Fig. 2-18. This circuit is called a bistable multivibrator, or *flip-flop.*

Each SCR, Q1 and Q2, drives its own independent load (the lamps). When one of the lamps is ON, the other will be OFF. Both lamps can never be simultaneously lit unless one of the SCRs is shorted out, or there is some other major circuit defect.

To understand how this circuit works, let's assume that initially SCR Q1 is ON and SCR Q2 is OFF. Under these circumstances, capacitor C1 is fully charged, with the end by Lamp 2 positive. This is a stable state for the circuit. It will hold this state as long as power is supplied to the circuit and neither of the pushbutton switches are depressed.

If switch S1 is closed while the circuit is in this state, nothing will happen. Another gate pulse will be fed to SCR Q1, but since this device is already ON, it will have no effect.

If switch S2 is momentarily closed, SCR Q2 will be turned ON and capacitor C1 will discharge, making the anode of SCR Q1 negative for a few microseconds. This switches OFF SCR Q1. Now, capacitor C1 will recharge with the opposite polarity. This time, the end by Lamp 1 will be positive. Again, this is a steady state for this circuit. This state will be held as long as power is supplied to the circuit and neither of the push-button switches are depressed. Closing switch S2 a second time will have no effect on the operation of the circuit.

If switch S1 is momentarily depressed (closed) while the circuit is in this state, SCR Q1 will turn ON and SCR Q2 will turn OFF. We'll be back in the original stable state described above.

Since this circuit can be switched between two stable states, it is called a bistable multivibrator. If both switches are simultaneously closed, it will be a disallowed state. The circuit state when the switches are released is unpredictable. It is possible that the SCRs or their loads could be damaged by such a disallowed state.

The circuits described in this chapter are for demonstration purposes only. They were presented here to illustrate the operation of an SCR. They are not intended as finished practical projects. For this reason, no parts lists have been supplied for these projects.

Experiment with various component values, according to the guidelines described in the text for each of these circuits. A number of useful and practical projects using SCRs and other types of thyristors will be presented in chapter 10.

❖ 3
Triacs

SCRS, LIKE DIODES, ARE UNIDIRECTIONAL, BUT YOU SHOULD BE AWARE THAT this is not the case for all types of thyristors. SCRs have a definite polarity. Current can only flow in one direction. If the input is an ac waveform, the SCR can conduct for only half of each cycle, or less. For some applications, these characteristics are desirable. For other applications, they are a major disadvantage.

When a given application requires bipolar current flow, an ordinary SCR just won't do the job. One possible solution would be to wire a pair of SCRs back-to-back, as illustrated in Fig. 3-1. Effectively, we now have a bidirectional SCR. Current can flow in either direction.

This trick is so useful, that semiconductor manufacturers make it unnecessary. They make self-contained specialized components which contain a pair of back-to-back SCRs in a single package. These devices are called *triacs*. The name is intimidating, but if you understand the SCR, you should have no problem at all comprehending and using the triac.

COMPARING TRIACS AND SCRS

The schematic symbol for a triac is shown in Fig. 3-2. Notice that this symbol suggests back-to-back SCRs. As with the schematic symbol for an SCR, the circle is optional. Its use or non-use is simply a matter of personal preference.

Like an SCR, a triac has three leads. The gate lead serves the same function as on an SCR. It controls the current flow through the main body of the device. The signal applied to the gate terminal is normally used to turn the triac ON.

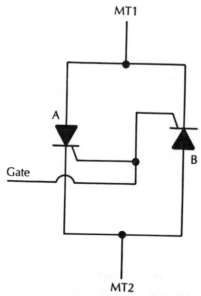

Fig. 3-1 *For bidirectional operation, a pair of SCRs may be wired back-to-back.*

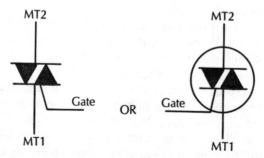

Fig. 3-2 *A triac is functionally the equivalent of a pair of back-to-back SCRs in a single housing.*

Since current can flow between the other two leads in either direction, the terms "anode" and "cathode" would be inappropriate. These leads on a triac are usually labeled "MT1" and "MT2." These terms are abbreviations for "Main Terminal 1" and "Main Terminal 2." For the vast majority of practical applications, these two leads are fully interchangeable, and the triac can be connected into a circuit in either direction.

When properly triggered, a triac can pass current in either direction. In a power controller circuit with an ac waveform as the input, a triac can pass anything from about 0 percent to almost 100 percent of the input signal. More of the input power can be made available at the output. A comparable circuit using an SCR instead of a triac can only pass from 0 to 50 percent of the input waveform, because an SCR permits current to flow in only one direction.

Triacs are pretty efficient devices overall. They are potentially at least twice as efficient as SCRs, depending on the specific operation of the circuit being used. The reason is that triacs are bidirectional, while SCRs are unidirectional. An SCR must always block half of every ac cycle. This means, that half of the input power is, by definition, wasted. A triac, on the other hand, can pass virtually all of the input cycle, with little inherent waste.

Of course, wasted power in a device such as an SCR or triac is dissipated as heat. While the use of a heatsink is always a good idea for many applications, you can get by with a fairly modest heatsink on a triac, even up to the recommended limits defined by the manufacturer's specification sheet for the individual unit used. When in doubt, use an external heatsink, just to be on the safe side. Too much heatsinking will never adversely affect circuit operation.

While efficient, triacs certainly are not perfect switches. There will be some voltage drop across the device. Typically one or two volts will be dropped across the triac, and this power will be dissipated as heat. Usually, some kind of external heatsink will be strongly advisable, but it does not have to be large or elaborate, because this is not a very large amount of power to be dissipated.

Triacs are often packaged in standard transistor housings, such as the small TO-92 (Fig. 3-3), the TO-220 (Fig. 3-4), and the rather hefty TO-48 (Fig. 3-5). As a general rule, devices rated for relatively low

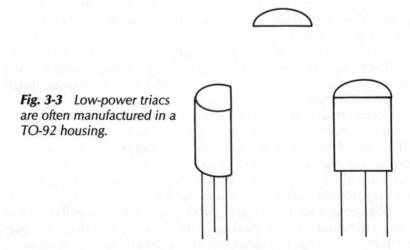

Fig. 3-3 *Low-power triacs are often manufactured in a TO-92 housing.*

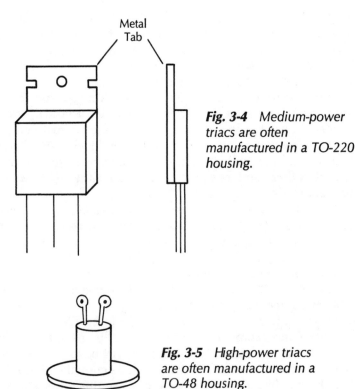

Metal
Tab

Fig. 3-4 *Medium-power triacs are often manufactured in a TO-220 housing.*

Fig. 3-5 *High-power triacs are often manufactured in a TO-48 housing.*

power dissipation use the TO-92 style housing, while heavy-duty triacs use the TO-48 case style. TO-220 cases are commonly used for triacs with in-between ratings.

The metal tab on the TO-220 housing and the metal stud base on the TO-48 housing serve as built-in mini heatsinks, although some external heatsinking will probably be required too, especially in high-power applications.

The metal tab or stud is often, though not invariably internally connected to the MT2 terminal. An insulating washer made of mica or Teflon should be placed between the metal tab on the triac and any external heatsink, metal case or chassis. This insulation is particularly vital in any circuit using ac line current. Without adequate insulation, a very dangerous potential shock hazard exists.

A few triacs have an insulated metal stud or tab. This will normally be noted in the manufacturer's specification sheet. If you're not 100 percent sure, it is wisest to assume that the metal tab or stud is not insulated. When in doubt, use external insulation, just to be on the safe side. Using an insulating mica or Teflon washer will never hurt anything, even if it is not strictly necessary. Why take needless chances?

INTERNAL CONSTRUCTION

The internal construction of a typical triac is illustrated in Fig. 3-6. Notice that this device is rather similar to a pnp transistor, except there is no internal difference between the collector and emitter terminals. Of course, being a thyristor, the triac actually has a pnpn structure, but the resemblance to a pnp transistor is still quite strong.

Compare this diagram to the internal structure of an SCR, as discussed in chapter 2.

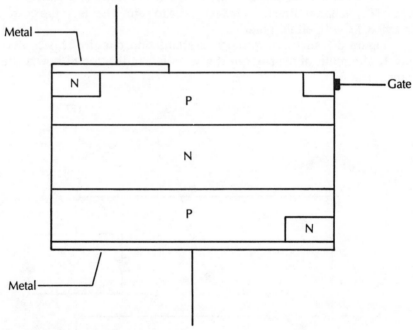

Fig. 3-6 *The internal construction of a triac (simplified).*

TRIAC OPERATION

Not surprisingly, a triac's operation is not very dissimilar to that of an SCR. The gate triggering function works in exactly the same way for both SCRs and triacs.

An SCR is turned OFF if the current flowing through it reverses polarity or drops close to zero. But a triac is supposed to conduct current in either direction. You may well be wondering just how we turn the thing OFF once a trigger pulse on the gate has turned it ON.

This is a little easier to understand if you remember that the triac is effectively equivalent to a pair of standard SCRs connected back to back. Refer to Fig. 3-1. If MT1 is positive with respect to MT2 when the triac is triggered at its gate, SCR A will be turned ON. SCR B will stay OFF, because its anode is negative with respect to its cathode. When the positive signal on MT1 drops below the holding current, I_h, SCR A will be switched OFF.

Of course, if MT2 is positive with respect to MT1, then the entire situation is reversed. Now, when the triac is gated, SCR B will be turned ON. SCR A will stay OFF, because its anode is negative with respect to its cathode. When the positive signal on MT1 drops below the holding current, I_h, SCR B will be switched OFF.

The triac will be cut OFF when the current flowing between MT1 and MT2, in either direction, is very close to zero. That is, if the current is below I_h, with either polarity.

Figure 3-7 shows a crude triac simulation circuit. This is not a practical circuit, of course, but it gives a fair indication of how a triac functions.

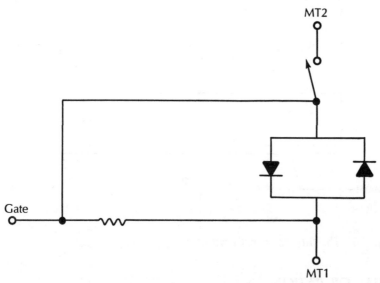

Fig. 3-7 *This circuit is a crude simulation of a triac.*

TRIGGERING MODES

Because of its bidirectional current carrying capabilities, the triac is a very versatile device. There are four possible combinations for triggering a triac. We call these the triggering modes. The differences between the modes lie in the relative polarities of the leads.

Normally, all current and voltage polarities for a triac are given with respect to MT1. This convention is followed, simply as a matter of convenience and to avoid unnecessary confusion. There is nothing particularly special about MT1. In fact, MT1 and MT2 are functionally identical. A triac is a nonpolarized device, and it is therefore symmetrical. It works the same way, forwards or backwards. MT1 is just an arbitrarily chosen common reference point.

The triac's four standard triggering modes are as follows:

- MODE A
  MT2 positive with respect to MT1
  Gate pulse positive with respect to MT1

- MODE B
  MT2 positive with respect to MT1
  Gate pulse negative with respect to MT1

- MODE C
  MT1 positive with respect to MT2
  Gate pulse positive with respect to MT1

- MODE D
  MT1 positive with respect to MT2
  Gate pulse negative with respect to MT1

Each of these triggering modes has a different current requirement to trigger the triac.

Mode A is the easiest to trigger. It has the lowest current requirement. The current at the gate required to trigger the triac in Mode A is I_{gt}.

In Mode B the triac isn't nearly as efficient. A gate current of at least five times I_{gt} is required to trigger the triac in Mode B.

Mode C and Mode D are basically similar to one another. A gate current of about twice I_{gt} is needed to trigger the triac in these modes, regardless of whether the gate is positive or negative with respect to the MT1 terminal.

Triacs, like all thyristors, or like all semiconductor components, are temperature sensitive. That is, the operating parameters of a triac vary with changes in its temperature.

The dc holding current, I_h, will normally decrease with increases in temperature, and vice versa. This is also true for the required gate trigger current I_{gt}, and gate trigger voltage, V_{gt}. These characteristics provide another good reason for the use of a heatsink.

Even if the triac doesn't overheat enough to cause damage to itself, it could easily run hot enough to change its own operating parameters. This could lead to erratic operation, and possibly damage to some other components in the circuit. A good circuit design does not leave anything to chance. All significant parameters are controlled by the circuit.

TRIAC DEMONSTRATION CIRCUITS

As was the case for chapter 2, the circuits presented here are for demonstration purposes only. They are shown here solely to illustrate the functioning of triacs. They are not intended as complete, practical projects, but as experimental demonstration circuits. Functional and practical projects for you to construct and use in specific applications are presented in chapter 10.

PROJECT 12: SIMPLE TRIAC DEMO CIRCUIT

The circuit shown in Fig. 3-8 is about as simple as it can possibly be. The load, or device to be controlled, is in series with the triac. The ac input line voltage is fed across the triac and load.

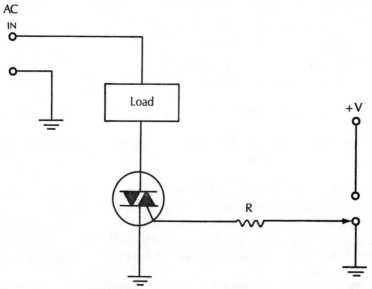

Fig. 3-8 *This is a super-simple triac demonstration circuit.*

The only other components in this circuit are a gate resistor and switch. In some cases, the resistor may not be necessary.

In a practical circuit, the switch would probably be replaced by an electrical signal. For our purposes here, a mechanical switch makes the concepts a little clearer.

Normally, the switch is in its lower position, as shown in Fig. 3-8. This keeps the gate terminal grounded. That is, there is no gate signal. Or, you could say that the gate signal has a value of zero volts. Before applying a gate signal, let's assume that MT1 is much more negative than MT2.

Now, if we briefly move the switch to its upper, V+, position, creating a gate pulse, the triac will be triggered. The triggering process takes a few microseconds. For most practical purposes, we can say that the triac turns on instantly. In some very critical, high-speed applications, the triac's switching time may be significant, but usually, it may be reasonably ignored without problems.

Like an SCR, once a triac has been turned ON, it stays ON, as long as current continues to flow between its main terminals, MT1 and MT2. When the input signal is a symmetrical ac waveform, such as a sine wave, the triac will turn itself OFF on each zero crossing of the input waveform.

In the circuit of Fig. 3-8, the gate voltage used to trigger the triac is assumed to be positive. This is arbitrary. The gate signal for a triac may actually be either positive or negative, depending on the requirements of the individual application. Refer back to the section on Triggering Modes earlier in this chapter.

PROJECT 13: SIMPLE AC POWER CONTROLLER

Figure 3-9 shows a very minor modification of the basic triac circuit of Fig. 3-8. Here the triac is serving as a lamp dimmer, controlling

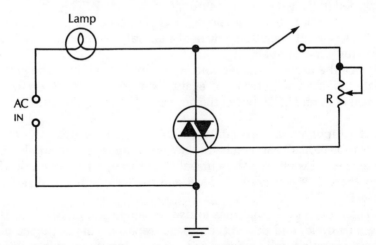

Fig. 3-9 *This minor variation on the circuit of Fig. 3-8 offers manual control over the output power level.*

how much power, or how much of each input ac cycle is fed to the light-bulb, which is the circuit's load. The value of resistance R controls the brightness of the lamp.

When the switch is open, nothing much happens, of course, because there is no signal being applied to the gate terminal of the triac. The triac is in its OFF state and it remains that way. The circuit is enabled by closing the switch. This allows the input voltage to reach the triac's gate via resistance R.

At the beginning of each cycle, the instantaneous voltage is zero. So zero volts are applied to the gate. The triac is OFF. If it was ON before, it has now switched OFF, because there is no current flowing between MT1 and MT2 at this point in the cycle.

Now, the instantaneous voltage and current of the input ac signal will start to rise. At some point, the triggering threshold of the triac will be exceeded at the gate. The value of resistance R determines how much of the input voltage will reach the gate. The larger this resistance is, the higher the input voltage must be (the further along in the cycle it must be) to exceed the device's trigger threshold.

When the triac's trigger threshold is exceeded at its gate, the unit switches ON, permitting the input current to flow between MT1 and MT2.

The cycle of the ac waveform at the input will continue up to its peak value, and then the voltage and current will start to drop. At some point, the current flow through the triac will fall below the holding current, I_h. The triac will be turned OFF. The input signal passes through the zero point and starts to go negative.

At some point, the negative trigger voltage will be exceeded by the voltage fed to the gate terminal through resistance R. The negative trigger voltage requirement will often be greater than the positive trigger voltage, so circuit operation may not be entirely symmetrical. This generally won't matter at all.

Once the negative trigger voltage has been reached at the gate terminal, the triac will turn ON again. Now current will again flow between MT1 and MT2, but in the opposite direction from the first half-cycle.

The input signal will continue going negative, until the cycle's negative peak has been reached. Now the input signal starts back up towards zero. Eventually, the current flow through the triac will drop below the holding current, I_h, level and the triac will turn itself back OFF again.

The input signal continues to rise, passing through zero and starting a new cycle, and this entire process repeats. This sequence of ON and OFF operation will continue as long as ac input power is applied to the circuit and the switch is closed.

If R has a fairly low resistance, most of the input signal will be passed through to the lamp, as illustrated in Fig. 3-10. Most of the input signal is fed to the output load (the lamp), so the lamp burns at near full brightness.

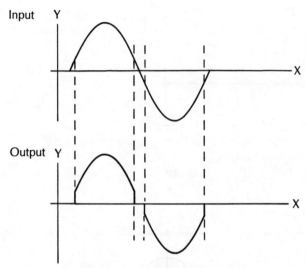

Fig. 3-10 *If R has a fairly low value, most of the input power will reach the output load.*

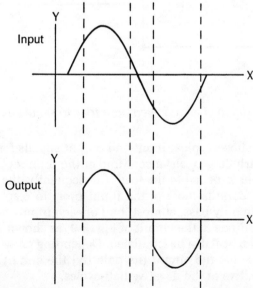

Fig. 3-11 *Typical input and output signals for the circuit of Fig. 3-9.*

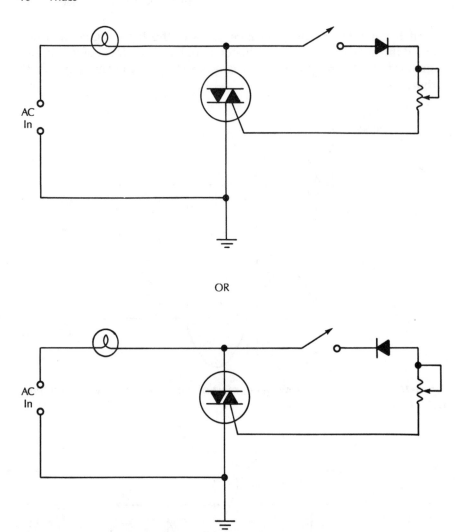

OR

Fig. 3-12   *The addition of a diode permits a triac to simulate an SCR.*

Figure 3-11 shows typical input and output signals for this circuit when R has a fairly large resistance. More of the input cycle is cut off, leaving less power to be fed to the load. Consequently, the lamp burns dimmer. A triac permits more of the input cycle to reach the output load. An SCR can only pass, at most, half of each input cycle.

By adding a diode to the circuit of Fig. 3-9, as shown in Fig. 3-12, the operation of an SCR can be simulated. Depending on which way the diode is inserted into the circuit (its polarity) the circuit can operate either on the positive or the negative half-cycles.

Naturally, a triac may be used in virtually any circuit that calls for an SCR. A diode may need to be added, as in Fig. 3-12, to force the triac to ignore alternate half-cycles of an ac waveform. This permits a triac to precisely simulate the operation of an SCR.

PROJECT 14: PHASE-SHIFT AC POWER CONTROLLER

The circuits presented in the preceding section are certainly more versatile than the SCR circuits of chapter 2, but they are still limited. In Fig. 3-9, the triac may be turned ON at any point between 0° and 90°, and between 180° and 270°. However, there is no way to turn the triac ON between 90° and 180° or between 270° and 360° using this simple circuit.

In electronics, there is usually some way to get around such limitations. In this case, the answer is to use a phase-shift network. This is a passive circuit made up of a resistor and a capacitor. Such a network is frequency and phase sensitive.

A simple triac circuit using a phase-shift network is illustrated in Fig. 3-13. By using the correct component values in the phase-shift network, the triac may be triggered at any point from 0° to 180° in the positive half-cycles and anywhere between 180° and 360° on negative half-cycles.

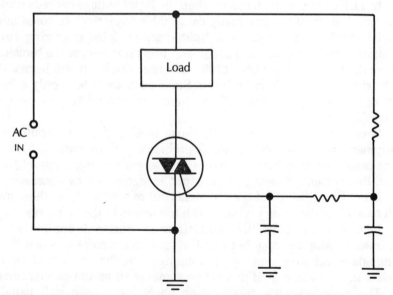

Fig. 3-13 *This triac circuit includes a phase-shift network for a wider triggering range.*

PROTECTING AGAINST TRANSIENTS

In the previous discussions, we have been assuming that the ac input power signal is a nice clean square-wave. In the real world, ac line current usually tends to be fairly noisy. Often, there are sharp transients on the power lines, resulting in an irregular waveform like the one shown in Fig. 3-14. Transient spikes can often be quite large in comparison to the nominal line voltage.

Fig. 3-14   *AC power lines often carry sharp transients and noise.*

In a triac circuit that triggers when the input voltage exceeds a specific point, a transient may easily cause false triggering. In some noncrucial applications, such as a light dimmer, false triggering from transients won't be much of a problem, unless it is severe. If a transient causes the triac to trigger ON a little early, then the lamp will burn a little brighter for that particular cycle. Since each cycle lasts only a fraction of a second, the effects of the transient aren't likely to even be noticeable.

Some applications, especially those involving any computerized equipment, could have considerable problems with transients. Also, in some cases, you may have to work with a severely noisy, frequent transient, line current. Filtering of the input ac signal may be necessary.

Even if we filter all of the transients out of the gate line, there may still be a false-triggering problem with transients in the ac power lines.

In most triac circuits, the incoming ac line current is applied across the triac, flowing between MT1 and MT2. If the current between these terminals rises high enough at a fast enough rate, the triac could switch itself to its ON state, even if there is no signal at all on the gate terminal.

The manufacturer's specification sheet for a triac will usually include a rating in the form of:

$$\Delta v/\Delta t$$

or:

$$dV/dT$$

The small triangle or the "d" in these terms represents the Greek letter "delta." This symbol is used to indicate a changing, rather than constant value. If the rated dV/dT value is exceeded, the triac will be triggered ON, regardless of the gate signal.

Essentially what is happening here is that the internal capacitance of the triac is being charged to a point greater than the device's nominal turn-on value. In ordinary operation, the internal capacitance is continuously charged, discharged, and recharged in the opposite direction, as the ac signal keeps reversing polarity. Normally, the charging of the internal capacitances involves negligible currents and can be entirely ignored. A sharp transient in the ac signal, however, can significantly increase the charging current.

The formula for the charging current is as follows:

$$I_c = C \times (dV/dT)$$

where C is the capacitance, (in farads), and dV/dT is the rate of voltage change in terms of volts-per-second. The charging current, I_c, is given in amperes for this formula.

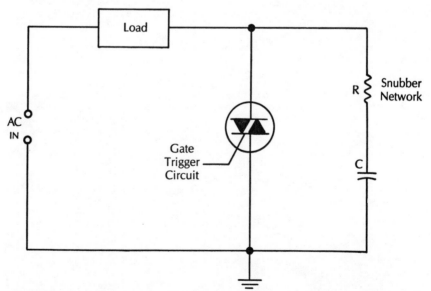

Fig. 3-15 *A snubber network helps a triac circuit filter out power line transients.*

If I_c exceeds the triac's gate triggering current, I_{gt}, the gate will be ignored and the device will switch into its ON state.

In critical applications, a special filter, or *snubber network*, consisting of an external resistor and capacitor should be placed across the triac, as illustrated in Fig. 3-15.

This snubber network is similar to the arc-suppression networks often found across the contacts in a relay circuit. The snubber network prevents false triggering of the triac due to sharp transients. The time constant of the snubber network, equal to RC, is selected for the specific application.

When power is first applied to the circuit, the snubber capacitor acts like a dead short to ground. After one RC time constant, the capacitor will be charged up to 63 percent of the supply voltage. After about five RC time constants, the voltage across the capacitor should be virtually identical to the actual supply voltage.

❖ 4
Diacs

IN THE LAST CHAPTER WE LOOKED AT TRIACS, WHICH ARE THYRISTORS WITH three leads. In this chapter we will examine a special two-lead thyristor called the *diac*.

WHAT IS A DIAC?

A diac is basically a specialized diode. Since the general definition of "diode" states that it is a two-lead semiconductor device, this should not be at all surprising.

Being a thyristor, the diac has a lot in common with the four-layer diode, discussed in the last part of chapter 1. However, there are some important differences between the diac and the four-layer diode. As you will soon see, the diac has a lot more in common with the triac.

Conceptually, a diac can be considered simply as a pair of back-to-back diodes, as illustrated in Fig. 4-1. Current of either polarity can flow through one of the two diodes; so the pair, as a unit, is bidirectional. Each diode is individually unidirectional, of course, as are all diodes. If an ac signal is applied across the pair, diode A will conduct during the positive half-cycles, while diode B will take over the conduction on the negative half-cycles. Each diode passes half the complete cycle. Together, they pass the entire waveform cycle. Because of this bidirectional conduction, you could call a diac a "two-way diode."

The back-to-back diode model is strongly suggested by the schematic symbols generally used to represent diacs, shown in Fig. 4-2. Like most other thyristor devices, the diac is used primarily for switching applications.

Being a switching device that can conduct current in either direction, the diac is often called a *bilateral switch*. The word "bilateral" in

Fig. 4-1 *A diac is like a pair of back-to-back diodes.*

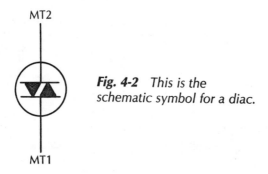

Fig. 4-2 *This is the schematic symbol for a diac.*

this context means that current may flow in either direction. The diac is a nonpolarized component.

THE DIAC AS A TRIGGER

In SCR ac power controller circuits, the SCR is often triggered via a neon lamp, as shown in Fig. 4-3. This triggering method may also be used with a triac, but the neon lamp can only trigger the thyristor (SCR or triac) during the rising portion of the positive half-cycle of the input waveform. In this case, the triac offers no particular advantage over the SCR. Both components will work in exactly the same way in circuits of this type.

If the neon lamp in Fig. 4-3 is replaced with a diac, as shown in Fig. 4-4, the triac will be bilaterally triggered. That is, it will be triggered on both positive half-cycles and negative half-cycles. This allows the circuit to operate more efficiently.

Diacs are usually, though not always, used in conjunction with triacs. This is because both diacs and triacs are bidirectional devices. An SCR, being an unidirectional device, cannot take much advantage of the diac's special capabilities. A diac may be used with an SCR, but in most applications there really isn't much reason to do so.

Consider the phase-shift controller circuit illustrated in Fig. 4-5. This circuit was discussed in chapter 3. While it might not be apparent, this circuit can be improved by adding a diac, as shown in Fig. 4-6.

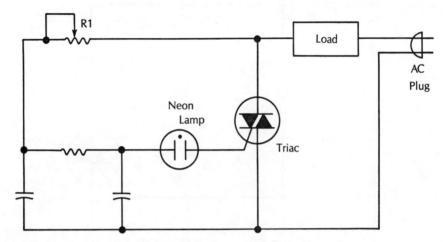

Fig. 4-3 *A neon lamp is often used to trigger an SCR.*

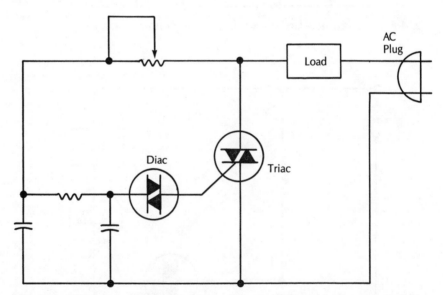

Fig. 4-4 *A diac can be used in place of the neon lamp in the circuit of Fig. 4-3 or bidirectional triggering.*

You should remember from chapter 3 that while a diac can be triggered with either a positive or negative gate pulse, a larger negative pulse is required to turn ON the device. This means that the triac will be turned ON later in each negative half-cycle than it is in each positive half-cycle.

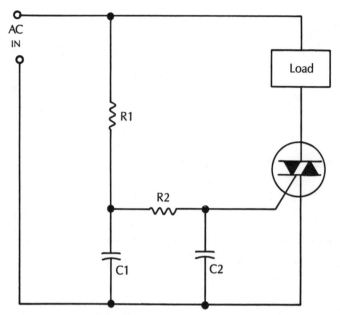

Fig. 4-5   *This is a standard phase-shift power controller circuit.*

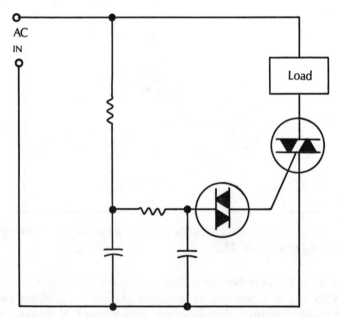

Fig. 4-6   *The phase-shift circuit of Fig. 4-5 can be improved by adding a diac.*

Using a diac neatly compensates for the differences in the positive and negative switching voltages required by the triac. The diac doesn't care at all if it is being triggered positively or negatively. In either case, it will put out a sharp pulse that is fed to the gate of the triac. The triac will switch on at approximately the same point in the negative half-cycles as it does in the positive half-cycles.

INTERNAL CONSTRUCTION

The internal construction of a typical diac is illustrated in Fig. 4-7. Compare this to the internal construction of a triac, which is shown again in Fig. 4-8. Notice how similar these two devices are to one another. In fact, there is only one real difference between these two devices: the triac has a third lead (the gate).

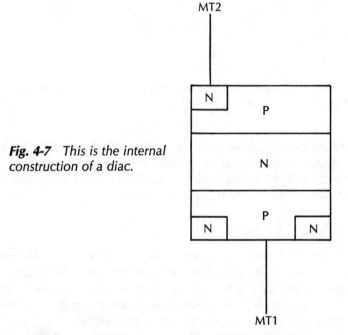

Fig. 4-7 *This is the internal construction of a diac.*

The diac is nothing more than a two-lead, gateless triac. The two remaining terminals on a diac are labeled in the same way as on a triac—MT1 and MT2. Sometimes the terminals are labeled "Anode 1," and "Anode 2," but I feel this is a little misleading.

There is no difference between these two terminals. They are entirely interchangeable. The diac may be wired into a circuit in either direction. It is not a polarized device. This makes it rather unusual as

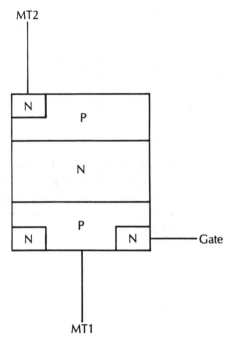

Fig. 4-8 *The only difference between the internal construction of a diac and that of a triac is the gate terminal.*

semiconductor components go. Most semiconductor devices have very definite polarity requirements, and can be accidentally wired into a circuit backwards. But it is impossible to hook up a diac backwards. It will work, no matter which way it is facing in the circuit.

OPERATION OF A DIAC

To understand how a diac works as a switch, take a look at the graph in Fig. 4-9. Normally, the diac is in its OFF state. Except for a small, and usually negligible leakage current, the diac does not conduct in its OFF state. For all intents and purposes, there is no current flow through the component under these conditions.

As the voltage applied across the diac is increased, at first, nothing happens. At some point, the breakover voltage, V_{bo}, will be exceeded. The exact V_{bo} value will depend on the specific characteristics of the individual device. This rating will be included in the manufacturer's specification sheet. For most diacs, the breakover voltage rating will be within the 20 to 40 volt range.

When the breakover voltage is exceeded, the diac switches to its ON state. It now starts to conduct current heavily. In its ON state, the diac exhibits negative resistance; that is, it does not obey Ohm's Law:

$$I = E/R$$

According to Ohm's Law, an increase in the current flow will result in an proportional increase in the voltage dropped across the component. This is ordinary positive resistance.

A negative resistance device, such as a diac in its ON state, works in just the opposite way. As the current flow is increased, the voltage drop is decreased. When a diac is switched ON, it produces a very large current spike, which can be used to trigger the gate of a triac or SCR.

A diac is turned OFF in a manner similar to that of SCRs and triacs. There is a specific minimum holding current, I_h. This value is listed in the manufacturer's data sheets. If the current flow through the diac drops below I_h, the device will switch back to its OFF state.

If an ac signal of a sufficient magnitude is fed across a diac, the diac will be repeatedly switched ON and OFF each half-cycle. A variable resistance element between the ac signal source and the diac can be used to vary at what point in each half-cycle the diac will turn ON.

Notice in Fig. 4-9 that the diac behaves in the same manner in both the positive quadrant and the negative quadrant. It does not care about

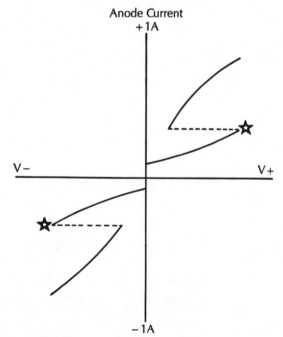

Fig. 4-9 *A diac functions as a bidirectional switch.*

the polarity of the voltage. A diac functions in the same way with both positive and negative signals.

ADDITIONAL APPLICATIONS

A diac can also be used as a variable ac resistance device. With a diac circuit, the output voltage can be regulated without loading effects from the load resistance. Even if the load resistance changes, the circuit's output voltage will remain constant.

❖5
Quadracs

FROM TIME TO TIME, MOST ELECTRONICS HOBBYISTS WILL ENCOUNTER SCRs, triacs and diacs. Actually, SCRs are becoming fairly common these days. Even triacs and diacs are being increasingly employed in many modern circuits.

But there is a close relative to these devices which most hobbyists are completely unfamiliar with. This is the *quadrac*.

While rarely used in hobbyist projects, quadracs are occasionally found in some commercial electronic equipment. It is useful to have some familiarity with this device, even though you may never use it yourself. So, this brief chapter will offer you a quick introduction to the quadrac.

INSIDE A QUADRAC

Since a diac has two leads (Di means two) and a triac has three leads (Tri means three), it might seem logical that a quadrac should have four leads, as Quad means four. Logical or not, this is not the case. A quadrac has only three leads, just like a triac.

I don't know the origin of the name, but I suspect it might be from the fact that a quadrac goes "one step further" than a triac.

The schematic symbol for a quadrac is shown in Fig. 5-1. If you've read chapters 3 and 4, you really already know just about everything there is to know about the quadrac.

As the schematic symbol suggests, a quadrac is basically a triac and a diac in a single housing. This is a very logical arrangement, because these two component types are frequently used together. The internal diac is hard-wired to the gate of the internal triac.

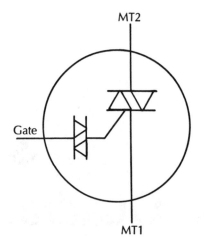

Fig. 5-1 *A quadrac consists of a diac and a triac in a single housing.*

Sometimes a quadrac is referred to as a "triac with trigger." The three leads to a quadrac are usually labeled the same as the leads for a triac. That is:

- Gate
- MT1
- MT2

In some sources, the main terminals, MT1 and MT2, may be called "anode 1" and "anode 2," or "high" and "common." Whatever names are used, it all amounts to the same thing. These terminals are exactly equivalent to the MT1 and MT2 terminals of a triac.

USING QUADRACS

A quadrac can be used in place of any triac and diac combination with no other changes in the external circuitry.

For example, take a look at the circuit shown in Fig. 5-2. This simple phase-shift power circuit has already been presented in several variations in the earlier chapters of this book. As shown here, the active elements are the diac and the triac. The diac is used to trigger the gate of the triac.

Figure 5-3 shows how the same circuit looks when the triac and diac are replaced by a quadrac. There really isn't much difference, is there?

Even though both active elements (diac and triac) are within a single housing, there is no difference in their functioning. The quadrac is an exact equivalent of a separate diac and triac. The only difference is

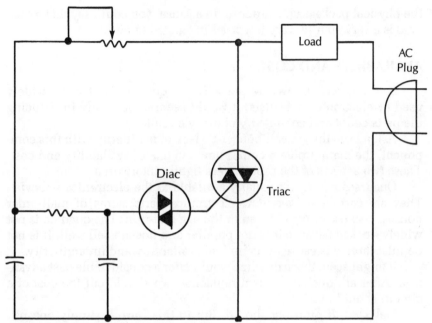

Fig. 5-2 *This simple phase-shift power controller circuit has been used throughout the earlier chapters.*

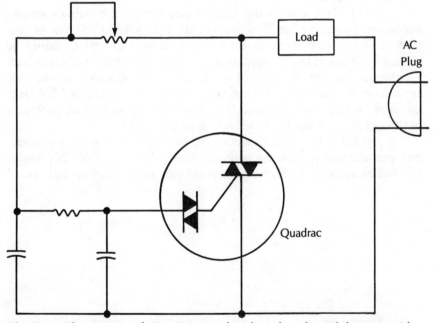

Fig. 5-3 *The circuit of Fig. 5-2 can be directly adapted for use with a quadrac.*

the physical packaging. I suppose, in a sense, you could say that a quadrac is a very simple, very low level integrated circuit.

AVAILABILITY AND COST

You may well be wondering why the quadrac isn't more widely used by electronics hobbyists. It would seem to be handy in reducing the parts count and complexity of many circuits.

Aside from the general hobbyist's lack of familiarity with this component, the main problem comes down to one of availability and cost. These two aspects of the problem are tightly interwoven.

Quadracs just aren't usually available to the electronics hobbyist. They are carried by very few electronics parts stores of mail-order houses, because of the chicken or the egg question. The quadrac is not widely stocked because it is not popular and doesn't sell well. It is not popular largely because of its limited availability and unfamiliarity.

It might seem that a quadrac would offer a considerable cost advantage. After all, one component replacing two should cut the cost of a circuit, shouldn't it?

Well, yes. It probably should. But in this case, it simply doesn't. Quadracs are sold for reasonable prices in large quantities, so they are appealing to manufacturers, especially in products where size is an important consideration.

Unfortunately, due to the limited popularity of quadracs among hobbyists, when you can find them at all, they tend to be quite expensive in single quantities. In most cases, it will almost certainly be cheaper and easier to use a separate triac and diac instead of a quadrac.

Because a quadrac doesn't offer that much of an advantage over separate triac/diac combinations, there isn't likely to be a sudden large demand for this type of device. The price versus availability problems of the quadrac are not likely to change in the near future.

If you really have your heart set on experimenting with a quadrac, one part number to look for is the T2006LT. You may find this device stocked by some of the larger mail-order electronics parts suppliers.

❖6
Dealing with Specifications

IF YOU ARE BUILDING A KIT, OR PUTTING TOGETHER A PROJECT FROM FINISHED plans, you can use the components supplied or specified by the circuit designer. But most electronics hobbyists go further sooner or later.

On the other hand, if you are designing a new circuit, or modifying an existing circuit, you will need to determine what components to use. People working with electronics also frequently need to find a suitable substitute for unavailable components.

It all comes down to specifications. What do all those numbers on the manufacturer's data sheet mean? Which specifications are truly important and critical, and which have more leeway? Which specifications can reasonably be ignored altogether in a given application?

In this chapter, we will examine the important specifications for thyristors. Under most circumstances, the most important specifications for thyristors are the PIV and power ratings. These specifications and several other operating parameters are discussed below.

PIV

The *Peak Inverse Voltage*, or *PIV* rating is one of the most important specifications for a thyristor. If you do not pay attention to this specification in choosing a thyristor for a specific application, the device could be damaged or destroyed. There is also a possibility of safety problems. (See chapter 8).

Anyone who works with electronics is probably already familiar with PIV ratings. This is exactly the same specification used for ordinary diodes.

The PIV rating is sometimes called *Peak Reverse Voltage*, or *PRV*. This alternate term isn't too frequently used for thyristors, although you may run across it now and then.

The Peak Inverse Voltage is the absolute maximum voltage that can safely be applied across a reverse-biased pn junction. If the PIV value is exceeded, the semiconductor junction will almost certainly be damaged.

The PIV rating of a thyristor should never be exceeded under any circumstances. If this rating is exceeded even momentarily, there is a very strong chance that the device will be damaged or destroyed. There is also a possibility of creating dangerous conditions if the PIV rating is exceeded.

It is vital to bear in mind that the PIV rating is a *peak*, not an *rms* value. Often, ac voltages are given in specification sheets and technical literature as rms values.

Rms, or root-mean-square, values are used to indicate the dc equivalent of an ac value. An ac voltage of 100 volts rms dissipates the same amount of heat through a given resistance as 100 volts dc. The peak value of an rms voltage will always be significantly greater than the rms value.

Ohm's Law relationships ($E = IR$) hold true for rms values, but not for peak values. This is why rms values are so useful and so widely employed for describing ac parameters.

Converting between rms and peak values can be complicated for many waveshapes. Fortunately, thyristors are used primarily for power control applications, and ac power voltages are usually in the form of sine waves. A sine wave is the simplest possible ac waveform.

For a sine wave, it is pretty easy to convert between rms and peak voltages. Just use the appropriate equation:

$$E_{peak} = 1.4\,E_{rms}$$
$$E_{rms} = 0.7\,E_{peak}$$

For example, the ac power lines in American homes nominally carry approximately 120 volts. This is an rms voltage. To find the peak value, we use the first of the formulas above:

$$
\begin{aligned}
E_{peak} &= 1.4\,E_{rms} \\
&= 1.4 \times 120 \\
&= 168 \text{ volts peak}
\end{aligned}
$$

To safely use a thyristor with ac line current, it should have a PIV rating of at least 168 volts.

You can also use these formulas to determine the maximum rms voltage that can be applied across a given thyristor. For instance, if you

have an SCR with a PIV rating of 400 volts, the maximum rms voltage
that can safely be applied across the component is:

$$E_{rms} = 0.7\, E_{peak}$$
$$= 0.7 \times 400$$
$$= 280 \text{ volts rms}$$

You probably won't be able to find a thyristor with a PIV rating
exactly equal to the calculated value. For instance, I have never encoun-
tered a triac with a PIV rating of 168 volts. This is just as well, because it
is always a good idea to overrate thyristors. Use a device rated for more
than the calculated value. The more "headroom" you leave, the safer
you will be. You can't use a "too high" PIV rating.

You could just use the largest PIV rating thyristor you can find in
every circuit, but this might easily become a case of major overkill. Thyristors with larger PIV ratings tend to be more expensive, and are often
quite large physically. If you go overboard on overrating thyristors you
will encounter the law of diminishing returns.

One major reason why overrating thyristors is advisable is that ac
power lines rarely carry pure sine waves. There will often be noise
spikes on the waveform. Sometimes these noise spikes can be quite
severe, as illustrated in Fig. 6-1.

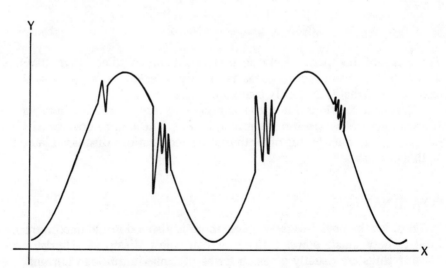

Fig. 6-1 AC power signals are often noisy and sometimes contain severe
spikes.

Ordinary ac line voltage has a peak voltage of 168, as we calculated earlier. But there could be noise spikes of 200 volts, or even more. Using an overrated thyristor will protect against such noise spikes.

Another possible solution to transient problems is to filter the line voltage before it is fed to the thyristor. In most practical thyristor circuits, both device overrating and filtering should generally be used, especially if the application is at all critical.

Here is a list of some popular SCRs and their PIV ratings:

IR122A	100 volts
IR122B	200 volts
IR122C	300 volts
IR122D	400 volts
C106D	400 volts
C116D	400 volts
2N3525	400 volts
BT109	500 volts
C126M	600 volts

Now here is a list of a few popular triacs and their PIV ratings:

C206D	400 volts
C226D	400 volts
SC146D	400 volts
TIC246D	400 volts
2N6073	400 volts

There is nothing particularly special about any of these thyristors. These devices were more or less randomly selected as fairly typical examples of what is currently available.

The specifications for these components are presented here as illustrations, not as specific recommendations. These same devices will be used as examples for the other thyristor specifications discussed later in this chapter.

POWER RATINGS

Besides the peak inverse voltage, there is also a definite maximum limit to how much power a thyristor can safely dissipate. Thyristor power ratings are usually given in terms of a maximum load current. Basic electronics theory tells us that power is the product of voltage and current:

Power (in watts) = Voltage (in volts) × Current (in amperes)

The maximum load-current rating assumes that the thyristor is being operated near its PIV rating. As a rule of thumb, use the following formula to calculate the maximum power that can safely be dissipated by any given device:

$$Power = I_{max} \times (0.8 \times E_{piv})$$

where I_{max} is the absolute maximum current flow the circuit will ever allow. Notice also that we drop the PIV value to 80 percent of its rated value to give us a "fudge factor" of extra safety.

The maximum current is often given as an rms, or average value. Use the following formulas to convert between rms or average, and peak values:

$$I_{peak} = 1.4 \times I_{rms}$$
$$I_{peak} = 1.57 \times I_{average}$$

$$I_{rms} = 0.7 \times I_{peak}$$
$$I_{rms} = 1.11 \times I_{average}$$

$$I_{average} = 0.636 \times I_{peak}$$
$$I_{average} = 0.9 \times I_{rms}$$

All of these formulas are correct only if the ac signal is in the form of a sine wave.

Always remember that all of these maximum values assume that a heatsink is being used on the thyristor. A large heatsink is not usually required, but for a component to dissipate the amount of power a thyristor is often expected to dissipate, some degree of heatsinking is almost always necessary. If you forego the use of a heatsink, make sure that the thyristor you use is amply overrated. Don't come anywhere near the maximum rated values without heatsinking the device. For more information on heatsinks, refer to chapter 7.

Here are the maximum current ratings for our sample thyristors. All of these current values are rms. First, for the SCRs:

IR122A	8 amperes
IR122B	8 amperes
IR122C	8 amperes
IR122D	8 amperes
C106D	4 amperes
C116D	8 amperes
2N3525	5 amperes
BT109	6.5 amperes
C126M	12 amperes

And, for the triacs, the power ratings are:

C206D	3 amperes
C226D	8 amperes
SC146D	10 amperes
TIC246D	15 amperes
2N6073	4 amperes

Once again, as with the PIV ratings, overrating the power of a thyristor is strongly advised. Using too large a thyristor will never hurt anything, although it may cost more and be more bulky. Overrate thyristors, but don't be ridiculous about it. There's no reason to use a triac rated for 12 amps in a circuit that will never draw more than 4 amps. You'll just be paying extra money for unused capabilities. Such overrating will not effect the electrical operation of the circuit in any way.

GATE SIGNALS

The manufacturer's specification sheet for a thyristor will also include maximum values for the gate signals. These are the maximum signal levels that will be required to reliably trigger the thyristor. These values may be exceeded. An input value less than the rated value may or may not be able to trigger the device.

There are two types of gate signals—the gate voltage (E_g) and the gate current (I_g). The following is a list of some typical gate specifications for our sample devices. First, the SCRs:

Device	E_g	I_g
IR122A	1.5 volts	25 mA
IR122B	1.5 volts	25 mA
IR122C	1.5 volts	25 mA
IR122D	1.5 volts	25 mA
C106D	0.8 volt	0.2 mA
C116D	1.5 volts	20 mA
2N3525	2 volts	15 mA
BT109	2 volts	3 mA
C126M	1.5 volts	30 mA

For some typical triacs, the gate specifications are:

C206D	2 volts	5 mA
C226D	2.5 volts	50 mA
SC146D	2.5 volts	50 mA
TIC246D	2.5 volts	50 mA
2N6073	2.5 volts	30 mA

As you can see, a large gate signal is not required for any of these thyristors. If you are substituting a thyristor in an existing circuit, make sure that the circuit provides a sufficient gate voltage/current to trigger the device. For example, if the existing circuit uses a C116D SCR, you may be able to substitute an IR122D SCR. Most of the important specifications for these two devices are similar, as the following comparison demonstrates:

	C116D	**IR122D**
• PIV	400 volts	400 volts
• Maximum Current	8 amps rms	8 amps rms
• E_{gt}	1.5 volts	1.5 volts
• I_{gt}	20 mA	25 mA

The IR122D can be substituted for the C116D as long as the circuit provides a trigger pulse at the gate of at least 25 mA. If the gate trigger pulse is only 22 mA, for example, the C116D SCR will work fine, but the IR122D probably won't.

To avoid damaging the thyristor, the gate trigger signal shouldn't be too large. Just use a little common sense. The gate trigger signal certainly shouldn't exceed the maximum power and PIV ratings for the component.

HOLDING CURRENT

An SCR or triac is turned OFF when the current flowing through it, between anode and cathode, or between MT1 and MT2, drops below a specific holding current value (I_h).

This turn-OFF value is determined by the internal construction of the specific thyristor used. It will be included in the manufacturer's specification sheet. For our sample SCRs, the I_h ratings are:

IR122A	30 mA
IR122B	30 mA
IR122C	30 mA
IR122D	30 mA
C106D	3 mA
C116D	35 mA
2N3525	20 mA
BT109	3 mA
C126M	35 mA

For our sample triacs, the I_h ratings are:

C206D	30 mA
C226D	60 mA
SC146D	75 mA
TIC246D	50 mA
2N6073	70 mA

When substituting thyristors, make sure that the holding current rating for the replacement device is reasonably compatible with the circuit. Generally, you have some leeway here. The exact I_h value is usually not too critical in most thyristor circuits.

For instance, the IR122D is rated for a holding current of 30 mA, and the C116D has an I_h rating of 35 mA. For the vast majority of thyristor circuits, these ratings may be considered equivalent. In a typical light-dimmer or power-controller circuit, the 5 mA difference probably won't be noticeable at all.

In addition, the I_h ratings are not 100 percent precise. There will be some variation between different devices of the same type number. Usually, the manufacturer's data sheet will list the maximum value for the holding current for the component.

❖ 7
Heatsinks

ALL SEMICONDUCTORS ARE INHERENTLY HEAT-SENSITIVE. TOO MUCH HEAT can easily damage or destroy a delicate semiconductor crystal.

Unfortunately, the very act of operating a semiconductor device generates heat within the component. Like all components in an electrical circuit, a semiconductor exhibits a certain resistance. A current flowing through a resistance results in a voltage drop, or loss, across the resistive element.

This "lost" voltage doesn't just magically go away. Energy can neither be created or destroyed. It can only be converted from one type to another. The electrical energy representing the voltage drop is converted into heat energy, which must be dissipated by the resistive element to the surrounding atmosphere.

In operation, a semiconductor could conceivably overheat itself enough to self-destruct if an excessive amount of current is being drawn through the device. Thyristors typically handle fairly large power levels, so heatsinks are usually strongly advisable.

Many thyristors, especially triacs, are extremely efficient. In many cases, only minimal heatsinking is required. Still, when in doubt, it never hurts to use a little extra heatsinking.

If a thyristor is being used well under its rated limits, you might be able to get away with using it without a heatsink. Still, you may be taking chances here; a good heatsink is relatively cheap insurance for a functional circuit.

HEATSINKING TABS

Many SCRs and triacs are housed in TO-220 cases, as shown in Fig. 7-1. The metal tab serves as a simple, low-grade heatsink. In some low-power circuits, the tab itself may be a sufficient heatsink.

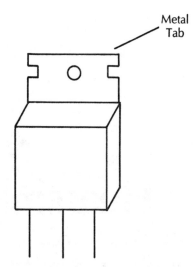

Metal
Tab

Fig. 7-1 *Many SCRs and*
triacs are housed in TO-220
cases.

Usually the tab will be used to connect the thyristor to a larger heat-sink, which can be any piece of metal exposed to air. It could be the chassis, or case housing the circuit.

In some thyristors, one of the leads is electrically connected to the tab. Many triacs have the MT2 terminal internally connected to the tab. This could be a problem in some circuits if you don't take it into account. Often the chassis or circuit housing will be grounded. Attaching the tab directly to a metal case or chassis could result in an unintended short to ground. Also, there is the possibility of some other conductor in the circuit accidentally touching the metallic heatsink. Again, the result is a short circuit.

To prevent such problems, you should install an insulating washer between the tab on the thyristor and the heatsink metal chassis, case, or whatever. The washer is usually made of mica, Teflon, or some similar nonconducting material. To ensure maximum heat transfer between the thyristor and the heatsink, the washer should be smeared with a special heatsink compound. Tubes of heatsink compound can be purchased from almost any electronics parts supplier.

Large power thyristors are often available in a TO-48 package, as shown in Fig. 7-2. This type of housing provides excellent heat transfer to the heatsink it is physically mounted on. The stud is put through a suitable hole in the heatsink metal, and is held in place with a washer and nut. In most cases, this stud will be electrically isolated from the thyristor's terminals because insulating this type of stud from the heat-sink is difficult.

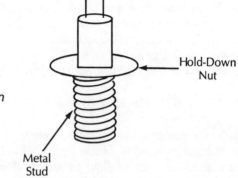

Fig. 7-2 *Large power thyristors are often housed in a TO-48 case.*

Hold-Down Nut

Metal Stud

SPECIALIZED HEATSINKS

A number of specialized metal heatsinks are commercially available. Some typical heatsinks of this type are illustrated in Fig. 7-3. The fins increase the metal surface-to-air contact area. This allows the heatsink to dissipate as much heat as possible.

Generally, the larger the heatsink, the more heat it will divert from the protected device. Fins make the heatsink effectively larger in a limited physical space.

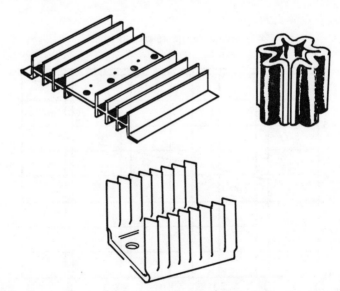

Fig. 7-3 *A number of specialized metal heatsinks are commercially available.*

Many of these heatsinks are designed to fit directly over the body of the thyristor or other semiconductor component. Others are designed to be screwed on via the device's tab, as discussed above.

Many commercial heatsinks are rated for a specific amount of power, usually given in watts. You can calculate how much power the heatsink will need to dissipate with this formula:

$$Power = (0.4 \times V_s)/8R_1 + (V_s \times I_d)$$

where the power value is given in watts. V_s is the maximum (peak) supply voltage in volts. R_1 is the device's load resistance in ohms, and I_d is the maximum current flow in amps. If you have some experience in metal working you should have no difficulty fabricating a heatsink of this type on your own.

PC BOARD HEATSINKS

Some thyristor circuits constructed on *printed circuit* (PC) boards can use copper traces on the PC board itself for heatsinking. If you know enough about the power to be dissipated in the circuit, you can calculate the area of copper needed for heatsinking. Some math is involved, but fortunately, the required calculations are not too complex.

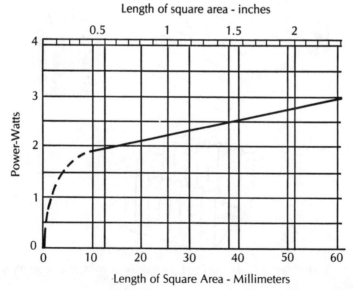

Fig. 7-4   *The required trace area for a PC board heatsink can be determined with the aid of this chart.*

The first step is to determine how much power will need to be dissipated. You can use this formula:

$$Power = (0.4 \times V_s)/8R_1 + (V_s \times I_d)$$

where the power value is given in watts. V_s is the maximum (peak) supply voltage in volts. R_1 is the device's load resistance in ohms, and I_d is the maximum current flow in amps.

The graph in Fig. 7-4 can be used to convert the necessary power dissipation into the required copper pad area for heatsinking. The copper pads do not necessarily have to be square or rectangular, but these shapes are usually the most convenient. Notice that this method of heatsinking is only practical for fairly low-powered circuits—typically three to four watts or less.

❖8

Safety Considerations

PLEASE READ THIS CHAPTER!

Some readers may be tempted to skip this chapter and just jump ahead to the projects. After all, safety is a pretty dry and mundane topic. You've probably heard it all before. This chapter may strike you as just a rehash of old hat material and common sense.

That may well be, but common sense is often a very rare commodity, especially when our minds are occupied elsewhere. Countless electronics hobbyists have needlessly destroyed expensive components, injured, or even killed themselves simply because they neglected a few simple safety precautions.

Please, do not give in to the temptation to skip this chapter. Read it carefully. We can all use a periodic reminder on safety procedures.

Unfortunately, all too many electronics hobbyists don't think too much about safety. It is too easy to get careless in the interest of efficiency and convenience. And sooner or later, the careless worker is going to get hurt.

Many thyristor circuits use ac line current. Carelessness here could result in a serious shock hazard, and possibly even death. No amount of efficiency and convenience is worth that kind of risk. Please don't take foolish chances.

It is very, very strongly advised that you read this chapter carefully, and consistently follow the recommendations given here.

FUSES AND CIRCUIT BREAKERS

In an ac powered circuit a fuse or circuit breaker should be used as a matter of course. There are no exceptions to this rule. Any time you omit a fuse or circuit breaker, you are setting the stage for a potential shock hazard, or a fire. You are also making it easy for your circuit to fail and for the components to be damaged or destroyed.

Fuses and circuit breakers are essentially automated switches used for circuit protection. When a dangerous power situation arises, the "switch" opens up, disconnecting the power source from the circuit before the problem condition can do any, or at least much, damage.

Fuses and circuit breakers are current sensitive. That is, they open up when the current drawn through then exceeds a specific value.

The voltage through a circuit can usually be reasonably controlled by the design of the power supply. Some kind of voltage regulation circuitry can be included if transients are likely to be a problem. In most cases, once the circuit is working correctly at the supplied voltage, you can reasonably assume that it will continue to do so.

But current is an entirely different story. The current flowing through a circuit is not determined by the power source, but by the circuit's own parameters. The current drawn through a circuit depends on the resistance and impedance (ac resistance) factors within the circuit itself. Remember Ohm's Law:

$$I = E/R$$

The voltage (E) will generally remain fairly constant, but the resistance (R) may change due to various conditions. This is especially true when ac signals are involved. Impedance can be influenced by a great many different factors.

According to Ohm's Law, as resistance drops in value, current increases. If the resistance decreases too much, an excessive amount of current may be drawn through the circuit.

As an extreme, but all too common example, consider what happens when there is a short circuit. When the circuit is working correctly (no short circuit) the resistance is probably moderate to high. When a short circuit forms, the resistance drops to a much lower value. In some cases, the short circuit resistance may be very close to zero. In such instances, the circuit will try to draw infinite, or at least very high current. Obviously, this is excessive.

Many electronic components, especially semiconductors, can be damaged by excessive current flow. If the current flow is high enough, it can damage or destroy any component.

In addition, this massive current flow will cause the components to get extremely hot. The combustion level of some materials may be reached by this heat. Fire is a very real risk.

Many short circuits may result in live ac current reaching something that could be touched by the circuit operator. His body will then serve as a conductor. If the current passing through his body is not restricted, electrocution could easily be the result. At the very least, the operator will receive a painful shock.

All this boils down to one thing—it is absolutely essential to limit the current flow through the circuit. Fuses and circuit breakers are an efficient way to do this. If the circuit tries to draw too much current, the fuse or circuit breaker will open and disconnect the power source from the load circuit.

Fuses

A fuse is really a very simple component. Basically, it is just a very thin wire. This wire is enclosed in a plastic, or metal housing to protect it from damage during handling. A typical fuse is illustrated in Fig. 8-1. Figure 8-2 shows the standard schematic symbol for a fuse.

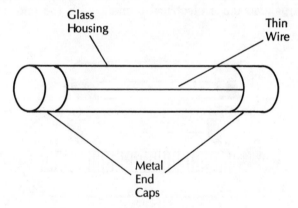

Fig. 8-1 *Fuses should be used in thyristor circuits for safety.*

Fig. 8-2 *This is the standard schematic symbol for a fuse.*

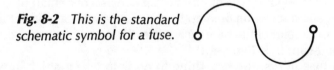

A wire acts like a low-value resistor. There is no such thing as a perfect conductor. As current flows through a wire, it inevitably heats up. If the current flow is high enough, the heat dissipated by the wire may actually exceed the melting point of the wire's material. The wire will melt, resulting in an open circuit.

A fuse is intentionally made from a wire with a relatively low melting temperature. When more than the rated current value flows through the fuse, the wire will heat itself up, and melt. The fuse will open and no more current will flow through it, or through the circuit it is protecting. Once the fuse wire has been melted, the fuse must be replaced. It is not reusable.

Sometimes fuses are soldered directly into a circuit, using "pigtail" leads. But since a blown fuse will need to be replaced, this is rather impractical in most applications, especially on the hobbyist level. Soldered fuses are occasionally used in some commercially manufactured equipment. The only reason I can think of for using a soldered fuse is to force users to take the equipment to a technician and pay a service charge for a simple blown fuse.

I think you can tell, I really don't approve of the idea. For convenient fuse replacement, some kind of socket is generally used. There are two types of fuse sockets in common use.

Most hobbyist projects use a pair of simple spring clips to hold the fuse in place. This type of fuse holder is illustrated in Fig. 8-3. The metallic clips also make electrical contact with the end caps on the fuse.

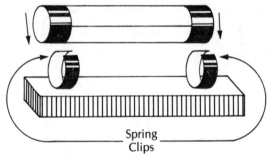

Spring
Clips

Fig. 8-3 *Spring clip fuse holders are handy.*

Another popular type of fuse holder consists of a small tube to hold the fuse, and a screw-on or snap-fit cap. One electrical contact is in the bottom, or far end of the tube receptacle. The other electrical contact is in the end-cap. (See Fig. 8-4.)

If a fuse blows, the first thing to do is to take a quick look at the circuit. If nothing is obviously wrong, replace the blown-out fuse with an exact duplicate. Sometimes a particularly severe transient will cause a fuse to blow, but nothing is really wrong. If the circuit now works without blowing the new fuse, then just forget about it. It was probably just an electrical fluke, and no damage was done.

If the new fuse immediately blows, or if it blows after just a minute or two of operation, then something is almost definitely wrong with your circuit. The odds against two fluke transients in a row are highly unlikely. Unplug the equipment and start troubleshooting.

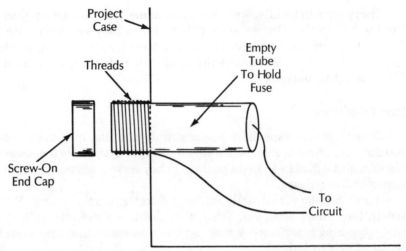

Fig. 8-4 *Another type of fuse holder uses a screw-on end cap.*

If the new fuse blows immediately, a short circuit is almost certainly the most likely culprit. Sometimes you will be able to find a short circuit by visual inspection. A loose wire or lead or even a loose glob of solder may possibly be making an electrical contact where it shouldn't. More frequently, a component has shorted out. You will need test equipment, such as a Volt-Ohm-Milliammeter (VOM), to locate the trouble.

If the circuit works for a short time before blowing the replacement fuse, look for a leaky component, most likely a capacitor, or a semiconductor, or a resistor which has somehow changed in value.

Don't automatically assume that just replacing the obviously defective component will always completely and permanently cure the problem. Take a careful look at the surrounding circuitry. Try to determine just what made that component go bad. Sometimes a component will just go bad on its own, but just as frequently, it's had some help from one or more other components or circuit conditions.

Never, never, replace a fuse with one of a higher rating. If one amp fuses keep blowing, don't even think about trying a 1.5 amp or 2 amp fuse. The circuit is supposed to work with its original fuse. If a fuse of that rating keeps blowing, then something is wrong. Ignoring the problem and using a larger fuse is dangerous. Some component in the circuit (probably an expensive semiconductor) will end up blowing to protect the fuse. I doubt if that's what you want.

I repeat, never, *under any circumstance, ever replace any fuse with one of a higher value.* Similarly, never bypass a fuse, even briefly for testing purposes. You're likely to end up doing more damage. *Don't do it.*

Every once in awhile, you hear about somebody putting a penny in the fusebox in their house wiring. Sometimes they get away with it. More often, some of their equipment is damaged. All too often, an electrical fire is started. Is it worth the risk? A fuse is there for a reason. Don't defeat its purpose.

Fused Resistors

Some circuits, especially power supply circuits, may use a fused resistor. The name of this device says it all. It is essentially a resistance element and a fuse in a single package. They serve a purpose similar to that of ordinary fuses.

Fused resistors are usually soldered directly into the circuit. Fortunately, they rarely blow out. When they do, it is a definite indication that something is seriously wrong, so it is reasonable that they must be replaced by a technician.

Circuit Breakers

Some equipment is protected by a circuit breaker instead of a fuse. A circuit breaker is more literally an automated switch. Unlike a fuse, a circuit breaker is reusable.

When a circuit breaker pops open due to a current overload, it can be reset by pushing in a small button. The button is usually red, and on most circuit breakers it physically moves outward when the circuit breaker is tripped.

If the circuit breaker repeatedly pops open, something is wrong with the protected circuit. Do not bypass the circuit breaker. This is the same situation as a circuit that keeps blowing fuses.

There is a slim possibility that the circuit breaker itself may be defective, but this is unlikely. When a circuit breaker does go bad, it will probably be impossible to reset it.

Resetting a circuit breaker is more convenient than replacing fuses. However, a circuit breaker is initially more expensive than a fuse and its holder. Also, circuit breakers tend to be physically bulkier than fuses.

Selecting Fuses and Circuit Breakers for Your Projects

In existing equipment, you know what replacement fuse to use. Just look at the value marked on the old one. But what if you are building, or even designing a circuit of your own. It's no trouble to find the current value for your fuse or circuit breaker.

First determine the maximum current that will ever be drawn by your circuit under normal operating conditions. Then just use the next higher standard fuse size. For example, if the circuit should never draw more than 2.87 amps, use a 3 amp fuse.

Before powering up the circuit, however, it is a good idea to double-check the manufacturer's specification sheets, at least for the more expensive components. Can they handle the maximum current permitted by the fuse? If not, replace them with heftier devices.

INSULATION

No conductor carrying a high voltage or current should ever be exposed, especially if the power is ac. It should not be possible for the operator to ever accidentally touch a live ac conductor.

Any circuit carrying ac power should be completely enclosed in either a metal or plastic case. If a plastic case is used, it should be impossible for any live conductor to make contact with the inside of the case.

Electrical shock is not fun. It can cause serious injury, or even death. At best it is very painful. Don't take unnecessary chances. Insulate your projects. This is absolutely vital for any circuit using ac power.

GROUNDING

Any circuit should be properly grounded. Again, this is of particular concern for ac power circuits, but it should not be ignored in dc circuits, especially when relatively large voltages or currents are involved. There are two types of ground; internal, or *common* ground and external, or *earth* ground.

The internal, or common circuit ground is usually necessary to make the circuit work. It is a common connection point for a return path to the power supply. Generally there should be a relatively large area common meeting point or ground plane. On a printed circuit board, a moderately large area of copper on the board can be used as the common ground point.

The earth ground is made externally to the circuit. Sometimes this type of ground is called "true ground." A conductor is run from the circuit to the earth itself. To ensure good electrical contact with the earth ground, a three to four foot metal spike is often driven into the ground. The ground wire from the circuit is then connected to this spike.

In some cases, an earth ground connection can be made to some pre-existing structural element, such as a cold water pipe. True earth grounds are usually not required for most modern semiconductor circuits, except when very large power levels or radio frequency (rf) signals are involved.

When the circuit is enclosed in a metal case, it is handy to use the conductive material of the case itself as the common ground point. Any connection made to any point on the case will be grounded.

When you use the case for grounding, it is especially important to do everything you can to prevent any possibility of any live conductor accidentally shorting to ground. Besides the shock hazard, this could easily damage or destroy many of the components in the circuit. Semiconductor devices are particularly susceptible to damage from such shorts. Component damage can occur even in a very low power dc circuit. Of course, the higher the power carried by the circuit, the greater the potential for damage.

In ac circuits, such a short could easily be a serious shock hazard. If you are using a grounded case for a circuit running from the ac power lines, use a polarized plug. A polarized plug has one prong slightly larger than the other. It can fit into the socket only one way. It can't be plugged in backwards. That way, the live power wire won't accidentally be connected to the case.

Better yet, use a three-pronged plug. The third, round prong is connected to a true earth ground within the socket. Connect the wire from this prong to the metal case to reduce the potential shock hazard to an absolute minimum. Even if there is a short circuit to the case within the project, the case will still be at ground potential. It will not be live if someone touches it. Of course, considerable damage may still be done to the circuit's components by such a short circuit, but at least no one will be injured or killed.

Safety is so important it doesn't make sense to ignore it. It is well worth the little bit of extra effort. Just try as much as possible to use that all too rare commodity—common sense.

❖ 9
Tips on Triggering

IN THE EARLIER CHAPTERS OF THIS BOOK, TRIGGERING HAS BEEN MENTIONED several times. Because this is such an important element in working with thyristors, triggering information is summarized in this chapter. Some, but certainly not all of the material in this chapter is repeated from elsewhere in the book. This chapter is intended as a handy reference source.

In much of this chapter, we will be primarily discussing SCRs, but much of this information applies just as well for triacs and quadracs too. Remember, triacs and quadracs are basically just "deluxe" bipolar SCRs.

THYRISTOR STATES

The SCR, like all thyristors, has two stable states. A stable state is one which can be held indefinitely. Unless the device is externally triggered, or power is removed from the circuit, the thyristor's current state will be held forever. This is equally true for both of its possible states.

Except for these two stable states, a thyristor normally has no other possible operating states. It is either in one stable state or the other. This is ignoring, of course, the brief transition times when the device switches from one stable state to the other.

The two stable states for thyristors are not too imaginatively named. They're pretty obvious really—just "ON" and "OFF." In the ON state, the thyristor conducts, and in the OFF state, it doesn't. It's really just as simple as that.

When a thyristor is ON, current can flow from the anode to the cathode, passing through a very low, almost negligible resistance. In the OFF state, the current flow is blocked by a very high resistance between the anode and the cathode.

Essentially, when the SCR is ON, it acts just like an ordinary forward-biased diode. When the SCR is OFF, it behaves like a reverse-biased diode, regardless of the polarity of the signal across the device.

GATE TRIGGERING

Normally, an SCR is triggered by applying an appropriate signal on its gate terminal. Later in this chapter we will consider some other ways in which a thyristor may be triggered. In the majority of thyristor circuits, a gate signal will be used. This is almost always the most practical and reliable method of triggering a thyristor.

If a voltage is applied between the anode and the cathode of an SCR nothing will happen as long as there is no signal on the gate lead. If a very low-level signal is fed to the gate, there won't be much of a reaction. The SCR will remain in its OFF state. The gate signal in this case is not strong enough to force the SCR to change states and switch itself ON. The SCR continues to block current flow from its cathode to its anode.

If the signal applied to the gate terminal exceeds the thyristor's threshold value, as defined by the manufacturer, the SCR will be triggered. It will switch into its ON state. Current can now flow through the device against only a small internal resistance, just as with an ordinary forward-biased diode.

This current will continue to flow through the thyristor, even if the voltage on the gate terminal is now removed. Once the SCR is switched ON, it stays ON, regardless of the gate signal. Retriggering the gate has no effect at all on the thyristor.

The only way to stop the current flow through the SCR, once it's been turned ON, is to drop the anode-cathode current to a point below the holding current (I_h) value. In simple terms, the gate can be used to turn the SCR ON, but it cannot turn it OFF again.

We are using the term "trigger voltage" here, because it seems to be the most logical name, although other terms are frequently used. Remember, on some spec sheets, the trigger voltage may be labeled "breakover voltage" (V_{bo}) or "switching voltage" (V_s). These terms are entirely interchangeable. Different manufacturers and technicians sometimes prefer different labels for certain parameters. It is sometimes confusing, but that's the way things are in the electronics industry.

The trigger pulse signal on the gate may be quite short, but remember that the SCR requires a finite amount of time to switch ON. Physics tells us that no action is ever truly instantaneous, and this definitely includes everything that happens in any practical electronic circuit. The gate pulse must last at least as long as the time it takes for the SCR to switch completely over to its ON state. Whether or not the gate pulse continues once the thyristor has been switched ON is entirely irrelevant.

As the SCR is switched ON, the anode current will rapidly increase until it reaches a specific latching value. This latching current value, (I_l), which is usually specified on the manufacturer's data sheet for the device, is generally somewhat higher than the holding current, (I_h). That is, once the SCR has been turned fully ON, the current level can drop a little without switching the device back OFF.

The latching current must be exceeded to turn the SCR ON. If the anode current drops below the holding current, the SCR will be turned OFF. This feature gives the thyristor some hysteresis.

The SCR requires a finite amount of time to switch on when it is triggered by a gate signal. For purposes of discussion here, this turn-ON time can be divided into two distinct stages: a delay time, (T_d) and rise time, (T_r).

The delay time is the time it takes the device to recognize and respond to the presence of a triggering signal on the gate terminal. The rise time is the period of time it takes for the SCR to actually switch ON once it "knows" it is supposed to. In more technical terms, the rise time is the time it takes for the anode current to rise from 10 to 90 percent of its maximum value.

In some circuits, especially where the load is inductive, such as a motor, or a relay, the gate signal may need to be held throughout the entire conduction period. Under certain conditions with an inductive load, the SCR may turn itself OFF prematurely if the gate signal is removed.

In the majority of applications, such problems will probably never occur. But you should be aware of their possibility, just in case you someday have to deal with a circuit with this kind of problem. Always use extra caution whenever you are using a thyristor circuit to drive any kind of inductive load.

In this discussion, we have been concentrating on SCRs. This was simply for convenience. Gate triggering works in exactly the same way for triacs and quadracs. The only difference is that these more sophisticated thyristor types can pass current in either direction when they have been properly triggered into the ON state.

POTENTIAL PROBLEMS

In any type of electronic circuit, there is always a possibility of problems cropping up, and thyristor circuits are certainly no exception. In this section we will look at a few common types of triggering problems to watch out for and what you can do about them.

In working with SCRs and other thyristors, you must remember that junction-leakage currents and current gain will increase with increases in temperature. The main effect of this is that the device can be triggered by a lower gate current when the ambient temperature is high than when the thyristor is in a relatively cool environment. The temperature reaction occurs both with both ambient (surrounding atmosphere) conditions, and self-heating effects due to the operation of the device itself.

While quite efficient, triacs are not perfect switches. There will always be some inevitable voltage drop across the device. Typically one or two volts will be dropped across the triac, and this power will be dissipated as heat. SCRs are generally less efficient than triacs, and so must dissipate greater amounts of waste heat. External heatsinking is often essential and always advisable in thyristor circuits.

During the turn-ON process, a high instantaneous circuit level will be dissipated through the SCR. If this power level increases too much or too fast, the SCR could be damaged or destroyed.

The manufacturer's specification sheet will include a dI/dT rating, which stands for a changing value.

Many of a thyristor's operating parameters may be affected by heat, whether ambient or self-generated. This is another good reason for use of a heatsink. Even if the thyristor doesn't overheat enough to cause damage to itself, it could easily run hot enough to change its own operating parameters. This could lead to erratic operation, and possibly damage to some other components in the circuit.

For example, the dc holding current, I_h, will decrease with increases in temperature, and vice versa. This is also true for the required gate trigger current, I_{gt} and gate trigger voltage, V_{gt}.

When a thyristor is first turned ON, it draws a very large current for a moment. The switching process takes considerable power and this high instantaneous power must be dissipated by the device.

To prevent the thyristor from self-destructing during turn-ON, the rate of increase of the anode current (dI/dT) must be limited to levels that can be safely handled by the specific thyristor used in the circuit.

The requirements for the gate signal are generally pretty flexible. As long as the gate signal is strong enough to trigger the thyristor, that's usually all you have to worry about. However, an excessive signal applied to the gate can do damage. Any semiconductor has its limits.

The manufacturer of the device will usually include the gate limit specifications in their data sheet. P_{gm} is the maximum power that can safely be dissipated by the gate terminal. Similarly, the maximum gate current is specified as I_{gm}. Also, the reverse voltage limit, V_{grm}, between the gate and the cathode must not be exceeded in operation of the circuit. Some manufacturers may use alternative names for some of these parameters on their spec sheets.

If the voltage on the anode is strongly negative with respect to the cathode, it is not advisable to apply a positive gate triggering voltage with respect to the cathode. This combination could easily force the SCR to dissipate an excessive amount of power and burn itself out.

One way to avoid such problems is to use a diode-resistor network, as illustrated in Fig. 9-1. When this network is connected between the gate and the anode, this reduces conduction between the gate and the cathode.

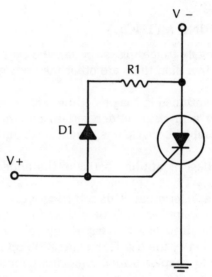

Fig. 9-1 *A diode–resistor network between the gate and the anode reduces the conduction between the gate and the cathode.*

A high leakage current can cause an SCR to be falsely triggered. A resistor wired between cathode and gate can help prevent this kind of problem. To make things easy on circuit designers, this resistor value will usually be included on the manufacturer's specification sheet.

An SCR may be triggered by a dc signal, but in some cases certain precautions may be necessary. The circuit should be designed so that it does not let the dc trigger signal to continue to reach the gate terminal during the reverse-blocking phase of the anode voltage—that is, when the SCR is reverse-biased.

If the dc gate signal is continuously applied, there could be excessive reverse leakage. The thyristor may be forced to dissipate a large amount of power, especially when high gate drive currents are used. In dc triggered circuits, some method should be included in the circuit to remove the gate signal during the reverse-blocking phase.

Thyristor data sheets generally list the latch current, I_l, and the holding current, I_h, with the assumption that the gate terminal is open. If the gate is forward-biased, these values will typically be reduced.

Similarly, reverse-biasing the gate will tend to increase the latch current and holding current levels.

In most circuits, these changes will be irrelevant and can be ignored. In some applications, the change in the latch current or holding current values could cause erratic or incorrect operation of the circuit. Such problems are most common when the gate is being driven by a small transistor with a saturation point of a few tenths of a volt.

OTHER TRIGGERING METHODS

Even though gate triggering is, by far, the best and most reliable way to turn ON a thyristor, there are other methods of thyristor switching.

As already mentioned in the preceding section, a high leakage current could trigger a thyristor under certain circumstances. This will usually be undesirable, because it is generally an uncontrollable circuit attribute.

If a large voltage is applied between the anode and cathode, or between MT1 and MT2, the thyristor will act like a diac, and switch ON, ignoring its own gate terminal. This self-triggering voltage will usually be specified in the manufacturer's data sheet for the device.

An SCR can be turned ON by a fast-rising voltage pulse. The necessary pulse is defined by the dV/dT specification on the manufacturer's data sheet. Refer to chapter 6 for more information about the dV/dT specification.

A voltage pulse like this could be intentionally applied to trigger the thyristor without using the gate terminal. It will do no damage to the thyristor. Essentially, in this method we are using the thyristor as if it were a diac.

In some other applications, however, triggering the thyristor via a dV/dT pulse may be undesirable for the intended circuit operation. False triggering of this type can usually be avoided, or at least limited, by connecting a capacitor between the cathode and the gate. A typical value for this capacitor is 0.05 μF.

While a fast voltage pulse usually won't damage a thyristor, unless it is very severe, a fast current pulse (Id/It) could be very harmful. Within certain limits, a fast current pulse could be used to switch ON a thyristor. But be careful. If the Id/It rating is exceeded, the thyristor could be damaged or destroyed.

To protect against problematic fast current pulses, you can add a simple series RC network between the anode and cathode, or between MT1 and MT2, as illustrated in Fig. 9-2. The component values should be selected to give an adequate time constant to sufficiently stretch out the anticipated undesirable fast current pulses.

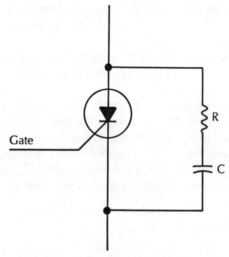

Fig. 9-2 *A simple series RC network can help protect a thyristor against fast current pulses.*

Since semiconductors are always temperature sensitive, a very hot environment could cause a thyristor to turn ON, even in the absence of any gate signal.

TRIGGERING TRIACS

The gate signal for a triac may be of either polarity. There are four possible combinations for triggering a triac. We call these the triggering modes. The differences between the modes lie in the relative polarities of the leads:

- Mode A
    MT2 positive with respect to MT1
    Gate pulse positive with respect to MT1

- Mode B
    MT2 positive with respect to MT1
    Gate pulse negative with respect to MT1

- Mode C
    MT2 negative with respect to MT1
    Gate pulse positive with respect to MT1

- Mode D
    MT2 negative with respect to MT1
    Gate pulse negative with respect to MT1

Normally, all current and voltage polarities for a triac are given with respect to MT1. Each of these triggering modes has a different current requirement to trigger the triac.

Mode A is the easiest to trigger. It has the lowest current requirement. The current at the gate required to trigger the triac in Mode A is I_{gt}.

In Mode B the triac isn't nearly as efficient. A gate current of at least 5 times I_{gt} is required to trigger the triac in Mode B.

Mode C and Mode D are similar. A gate current of about twice I_{gt} is needed to trigger the triac in these modes, regardless of whether the gate is positive or negative with respect to the MT1 terminal. A quadrac features the same four triggering modes as a triac.

❖ 10
Applications

ALL RIGHT, ENOUGH THEORY FOR NOW. LET'S PUT SOME THYRISTORS TO WORK in some practical projects. I'm sure many, if not most, of my readers will consider this the most important and interesting chapter of the entire book. After all, practical projects are almost always a lot more fun than theory.

To include as many thyristor projects as possible, this is by far the longest chapter in this book. Since this book assumes that you have some prior experience in electronics, detailed construction instructions are not given here.

The parts lists for these projects are as complete as possible. In many cases, however, some component values will depend on your specific application or on the load you intend to control with the thyristor circuit.

Most of the projects in this chapter are not presented in any particular order. Pick and choose from among these assorted projects according to your individual preferences and needs.

Every attempt has been made to offer as wide a variety of thyristor projects as possible. You are certainly encouraged to experiment with these circuits and modify them to suit new applications.

PROJECT 15: SCR TESTER

This first project doesn't actually include any thyristors in its circuitry, but I still think it is appropriate for this book. If you work with SCRs, sooner or later you will have to deal with some questionable units. This is true, of course, for any type of electronic component.

If you ever find yourself asking, "Is this SCR good or bad?" then you'll want this project. It is an SCR tester. The schematic diagram for

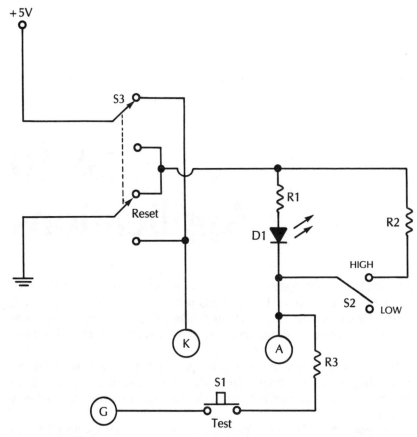

Fig. 10-1   *SCR Tester*

Table 10-1   Parts List for SCR Tester

D1	LED
R1	220 ohm $^1/_2$ watt resistor
R2	22 ohm 5 watt resistor
R3	100 ohm $^1/_2$ watt resistor*
S1	Normally open SPST pushbutton switch
S2	SPST switch*
S3	DPDT switch*

*—see text

this project appears in Fig. 10-1. The parts list is given in Table 10-1. As you can see, this is a fairly simple circuit.

This simple SCR tester does not read out any operating parameters of an SCR. It will not allow you to identify an unmarked and unknown device. The SCR tester is designed to perform only simple "Go/No-Go" tests of SCRs. Either the SCR is good or bad. The tester won't give you any more information beyond that. Fortunately, in a great many cases, especially when troubleshooting an existing circuit, this limited information will be quite sufficient.

The points in the circuit labeled "A," "K," and "G" are to be connected to the three leads of the SCR you want to test. Of course, the test points should be connected to the SCR's terminals in this order:

A   Anode
K   Cathode
G   Gate

Be very careful making these connections to the SCR being tested. Obviously, if the SCR's terminals are connected to the wrong test points, the test results will be completely meaningless.

You can use either a transistor socket, or you can attach small alligator clips to the ends of three test-leads. Personally, I think the alligator clip/test-lead approach is almost always the best choice for a component tester like this, because it will permit you to test SCRs in a wider variety of case styles.

This tester is not intended for in-circuit testing, although it can be used in that way in some cases. If you make in-circuit tests with this project, you really should take any test results with a large grain of salt. Other components in the circuit could easily cause an incorrect test result.

If the device tests good in-circuit, it is probably OK, but a bad test result really doesn't tell you much in this case. Other components in the circuit could be confusing the tester's readings. For the most reliable results, remove the SCR from its circuit before testing it.

When performing general troubleshooting, you could make your preliminary tests in-circuit. If the SCR tests OK, you can probably assume it is fine, and leave it alone. Look for the problem elsewhere in the circuit.

On the other hand, if the SCR tests bad in-circuit, remove it from the circuit and test it again in isolation. If the SCR tests OK, and you can't find any other problems in the circuit, you may want to desolder the thyristor and test it out of circuit, just to be sure.

In any event, making preliminary tests in-circuit can eliminate much of the fuss and bother of desoldering and resoldering component leads.

While the power supply for this project is shown as + 5 volts, this is not too critical. Five volts is a convenient size for an ac driven power supply, because many inexpensive five volt regulator ICs are available. If you prefer, you could use a six volt battery, such as four AA cells in series. The supply voltage for this project is quite flexible.

The more stable the supply voltage is, the better the circuit will operate. A regulated power supply is recommended for the best and most reliable results from this circuit. Now, let's take a moment to explore the actual circuit operation of this project.

Looking at the circuit shown in Fig. 10-1, switch S2 sets the test signal level. For low-power devices, open this switch to the LOW setting. For high-power units, S2 should be closed to its HIGH setting. The majority of thyristors probably will be tested at the HIGH setting, so you might also eliminate this switch, and hard-wire the connection to the HIGH point. However, you need to remember that there are some low-power SCRs on the market. Use the LOW setting of switch S2 for any thyristor rated for 200 mA or less.

S1 is a Normally-Open SPST push-switch. Closing this switch tests the SCR connected to the test leads by placing a signal on the gate terminal. If the SCR is good, the LED, D1, should light up. You might want to experiment with other values of resistor R3 to change the trigger voltage used in the test process.

If the LED does not come on in the LOW mode, try setting switch S2, to HIGH. If it still doesn't work, the SCR must be bad. Discard it and try a different unit. Note that in-circuit tests may sometimes give a false "bad" reading for a good thyristor.

Switch S3 will normally be left in the position shown in the diagram. Briefly move it to the opposite position to reset (turn off) the SCR after it is tested. A momentary-action switch would be suitable here. This switch is used to reset the SCR. Whenever the tester is reset via switch S3, the LED should go out.

If the LED comes on before you have depressed test switch S1, or if it stays on when the reset switch S3 is depressed, there may be a problem. It could just be a voltage glitch which has "fooled" the SCR into switching ON. Reset the device with switch S3. When switch S3 is returned to its normal test position, the LED should be dark. If it is still lit, the SCR is probably shorted. Discard it and use a new unit.

Triacs and quadracs may also be tested with this circuit. To check their dual polarity operation, each triac and quadrac should be tested twice. First, connect the MT1 lead to point A and the MT2 lead to point K and test the device. If everything tests OK in this position, reverse these connections. Now MT1 should be connected to K and MT2 should be connected to A.

A triac or quadrac should give the exact same test results in each of the two positions. This reversal of terminal orientation is necessary to fully test the bipolar operation of the triac or quadrac.

PROJECT 16: REMOTE POWER CONTROLLER

The circuit shown in Fig. 10-2 permits you to turn any ac load on and off from a remote location. A typical parts list for this project is given in Table 10-2.

You may need to change the parts values for various applications. Using the components listed in Table 10-2, the load may be up to 500 watts. If you change any parts values, try to provide more load capability than the intended load actually needs.

A little extra headroom is always a good idea. If the control circuit cannot adequately handle the requirements of the load, either the controller, or the load device, or both, may be damaged or destroyed. In addition, dangerous conditions, such as a shock hazard or a potential fire, could result from overloading the controller.

If you can't find the specific triac, Q1, called for in the parts list, substitute any available triac that is suitable for use with the intended load. Refer to chapter 6 for details on thyristor specifications.

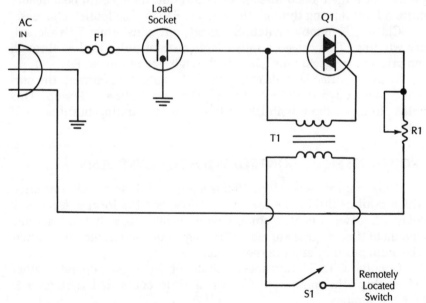

Fig. 10-2 *Remote Power Controller*

Table 10-2 Parts List for Remote Power Controller

Q1	Triac (GE-X12, or similar*)
T1	Transformer—secondary = 6.3 volts, 1 amp
F1	5-amp fuse (or to suit load)
R1	50-ohm, 2-watt potentiometer
S1	SPST switch

*—see text

Triac Q1 is triggered via the primary winding of transformer T1. Potentiometer R1 sets the trigger point within the ac power cycle. Carefully adjust R1 to avoid false triggering when the remote switch, S1, is open. Set the potentiometer for its maximum value without false triggering.

Switch S1 is any type of SPST switch mounted at a remote location. A doorbell type switch is a good choice for many applications. Twin-lead (two conductor) cable must be run between the remote switch and the control circuit.

This connecting cable should be well insulated to prevent any possibility of a shock hazard. Normally, these wires won't carry much power, but if the transformer should short out, there could be a deadly current level flowing through these wires. Don't take foolish chances.

Closing the remote switch, S1, shorts out transformer T1's secondary winding. This causes a rather large current pulse to flow through the primary, triggering the triac Q1 through potentiometer R1.

While the triac is in its ON mode and conducting current, the current flow through the transformer's primary winding will stop. This helps eliminate the possibility of the primary winding burning itself out.

PROJECT 17: PHASE CONTROL WITHOUT HYSTERESIS

The circuit shown in Fig. 10-3 is a phase-shift ac power controller, which exhibits little or no hysteresis. Hysteresis is a form of electronic delay. All devices exhibit some degree of hysteresis. It takes a finite amount of time for them to react after they have been triggered or otherwise influenced by any external signal.

Circuits with significant amounts of hysteresis operate rather "sluggishly." In some applications, a little controlled hysteresis is highly desirable.

Hysteresis tends to filter out overresponsiveness to any noise in the signal. On the other hand, hysteresis slows down circuit response, so it

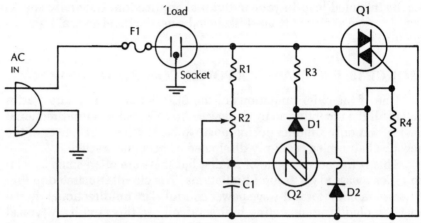

Fig. 10-3 *Hysteresis-Free Phase Controller*

Table 10-3   Parts List for Hysteresis-Free Phase Controller

Q1	Triac (SC40B, or similar*)
Q2	SBS (2N4992, or similar*)
D1, D2	silicon diodes (almost any)
C1	0.22 μF capacitor
R1, R3	4.7 K ¹/₂ watt resistor
R2	500K potentiometer
R4	100 ohm ¹/₂ watt resistor
F1	Fuse to suit load

*—see text

can be quite undesirable in certain applications, where an immediate response may be expected or needed.

The circuit illustrated in Fig. 10-3 is designed specifically to minimize hysteresis effects. This is done by resetting the capacitor C1 to zero at each zero-crossing after positive half-cycles.

Q2 is a Silicon Bilateral Switch or "SBS." This type of component is discussed in chapter 12. Potentiometer R2 is used to set the trigger point during the ac power cycles.

Of course, you may use a different triac (Q1) in place of the one called for in the parts list. This is just a typical unit for this application. If you substitute a different triac for Q1, check the manufacturer's specification sheet to make sure that it can handle the power levels needed for the intended load in your individual application. Generally speaking, this type of circuit is most likely to be used with electrical lights as the load.

PROJECT 18: FULL-WAVE SCR POWER CONTROLLER

One of the chief limitations of the SCR is that it normally wastes half of all the incoming ac input power. An SCR, being a unidirectional device, can only conduct during positive half-cycles. The power of the negative half-cycles is simply dissipated as heat and wasted.

This is the normal state of affairs, but there are often some special and tricky ways to get around limitations. The circuit illustrated in Fig. 10-4 uses an SCR for full-wave power control. The unidirectional thyristor is "fooled" into operating bidirectionally in this circuit. A typical parts list for this project is given in Table 10-4.

You'll notice that this parts list is rather incomplete. Many of the components must be individually selected to suit the specific intended application. A general-purpose parts list would be inappropriate in this case.

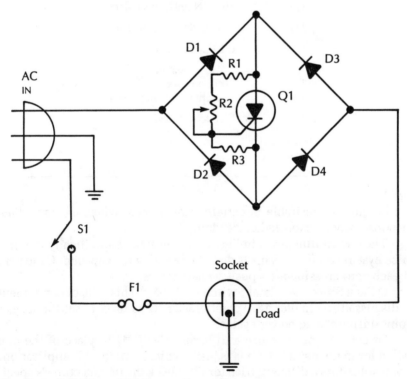

Fig. 10-4   *Full-Wave SCR Power Controller*

Table 10-4 Parts List for Full-Wave SCR Power Controller

Q1	SCR (C106B, or similar*)
D1 - D4	silicon diode—rated to match SCR (Q1)*
R1	100 ohm $1/2$ watt resistor
R2	1K 2 watt potentiometer
R3	100 ohm $1/2$ watt resistor
F1	Fuse to suit load

*—see text

Choose an SCR that can safely handle the absolute maximum power levels required by your intended load. The four diodes should have current and voltage ratings similar to the SCR. We'll get to resistor R1 in a minute.

The SCR is placed in a balanced bridge configuration with the diodes. A simple resistive voltage dropping network comprised of resistors R1, R2 and R3 determines the trigger point for the SCR. The diode bridge rectifies the input signal, so instead of a positive half-cycle and a negative half-cycle, the SCR sees two positive half-cycles making up each complete input cycle.

There is one important restriction for operating this circuit, which you should be aware of. This circuit is intended for purely resistive loads only. Do not use this project to drive inductive loads such as motors or relays. This control circuit will not work reliably with an inductive load, and the thyristor and/or the load might be damaged.

The value of resistor R1 should be selected so that the current in the load is at its required minimum level when potentiometer R2 is set for its minimum resistance. The correct resistance for R1 may be found experimentally. You could use a trimpot for R1. Usually a 1K or 500 ohm trimpot would be a good choice.

When this trimpot is properly adjusted, it can be carefully removed from the circuit, measured, and replaced with a fixed resistor of the measured value. Or, you could leave the trimpot in the circuit. It is often a good idea to use a drop of paint or glue to fix the trimpot's slider in the correct position, otherwise you might have to periodically recalibrate the circuit.

PROJECT 19: BASIC TRIAC LIGHT DIMMER

Probably the most common application for thyristors is in light dimmers. You will find several variations on the basic light dimmer in this chapter.

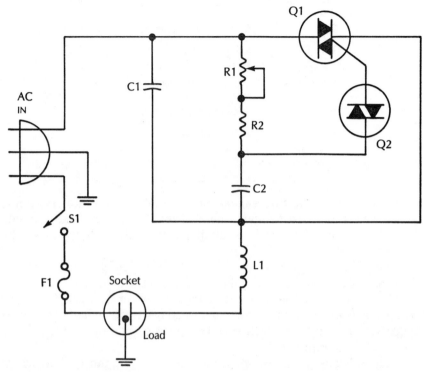

Fig. 10-5 *Basic Triac Light Dimmer*

Figure 10-5 shows a basic light dimmer circuit using triac Q1. A typical parts list for this project is given in Table 10-5. Of course, you can substitute other component values to suit the load in your individual application.

Triac Q1 and diac Q2 listed here are intended as typical examples. This circuit will work for almost any diac and triac combination. As a matter of fact, while it is rather unlikely, if you do happen to have a quadrac handy, you could use it in place of Q1 and Q2, providing its power ratings suit the intended load.

Adjusting potentiometer R1 determines exactly when during each power cycle the diac will fire and trigger the triac. This, of course, determines how much of the input power will be fed out to the load.

This circuit is perfectly straightforward; however, there are a couple of points we should make about it to set up the premise of the next project (No. 20).

This light dimmer project uses a single time-constant trigger. The triggering time is determined by the time constant of the combined series value of resistors R1, R2, and capacitor C2. This approach is quite

Table 10-5   Parts List for Basic Triac Light Dimmer

Q1	Triac (T2800B, or similar*)
Q2	Diac (D3202U, or similar*)
C1, C2	0.1 µF 250 volt capacitor
L1	100 µH coil
R1	250K potentiometer
R2	3.3K ½ watt resistor
F1	Fuse to suit load

*—see text

simple and straightforward, but it introduces a definite amount of *hysteresis*, or delay into the circuit's operation.

In this case, hysteresis means that there will be a noticeable difference in the potentiometer setting required to turn on the light from full off, and the setting at which the light will turn back off. For noncritical applications this won't be much more than a minor annoyance. But in some cases, the hysteresis effects can be a significant disadvantage. In such applications, the following project might be a more suitable choice.

PROJECT 20: IMPROVED TRIAC LIGHT DIMMER

The triac light dimmer circuit shown in Fig. 10-6 exhibits considerably less hysteresis than the one illustrated in Fig. 10-5. This feature also has a side benefit of effectively increasing the potentiometer's operating range. A typical parts list for this improved triac light dimmer project appears as Table 10-6.

The lower hysteresis from this project is due to the use of a double time-constant trigger network. The big difference between this circuit and the one used in Project 19 is the addition here of resistor R3 and capacitor C3.

The added capacitor charges to a higher voltage, thereby reducing the circuit's hysteresis. When the diac, Q2, fires, triggering the gate of triac Q1, the gate current pulse is obtained by the rapid discharge of capacitor C3. Capacitor C2 has a longer time constant than capacitor C3. This means when C3 is discharged, C2 still has some charge left.

The charge on capacitor C2 replaces some of the charge removed from capacitor C3 by the gate current pulse. This gives C3 a "head-start" on recharging for the next pulse. Essentially, C3's response is speeded up, decreasing the hysteresis, or delay, in the operation of the circuit.

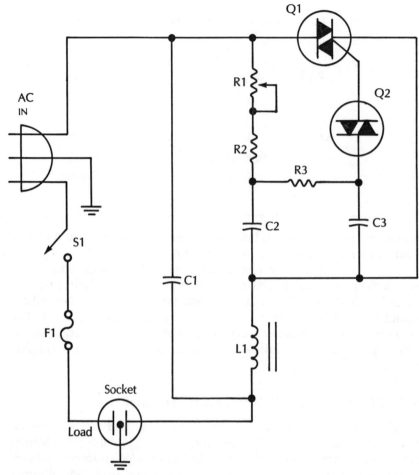

Fig. 10-6 *Improved Triac Light Dimmer*

Table 10-6 Parts List for Improved Triac Light Dimmer

Q1	Triac (T2800B, or similar*)
Q2	Diac (D3202U, or similar*)
C1, C2, C3	0.1 μF 250 volt capacitor
L1	100 μH coil
R1	100K potentiometer
R2	1K 1/2 watt resistor
R3	100 ohm 1/2 watt resistor
F1	Fuse to suit load

*—see text

PROJECT 21: LIGHT CROSS-FADER

The circuit shown in Fig. 10-7 is a kind of "dual light dimmer," with two separately controlled output loads. For this circuit, the loads will almost always be lighting devices of some kind.

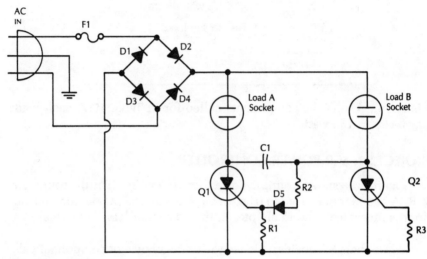

Fig. 10-7 *Light Cross-Fader*

The two loads are operated in a complementary fashion. As the power to load *A* is increased, the power to load *B* is proportionately decreased, and vice versa. If one of the loads is fully ON, then the other load will be fully OFF. In other words, this circuit permits a smooth cross-fade from one device to the other.

SCR Q1 operates pretty much in the usual way. When Q1 is in its OFF state, a small current will flow through load *A*, D1 and R1, charging up capacitor C1 and triggering the second SCR, Q2, ON. When Q1 is switched ON, the current will stop flowing through R1 and D1. Capacitor C1 will discharge, producing a negative spike, turning Q2 OFF.

This sequence of events will be repeated on each input cycle. The longer Q1 is held ON per cycle, the longer Q2 will be OFF, and vice versa. Q1 cannot supply more power to load *A* without decreasing the power Q2 can supply to load *B*. The two loads will always be perfectly balanced.

The parts list for this clever little circuit is given in Table 10-7. As always, you may substitute a different type SCR for Q1 and Q2, provided the replacement devices can supply sufficient power to the loads.

The two SCRs should be identical types to keep the circuit in balance. Other than that, there are no special restrictions involved in using

Table 10-7   Parts List for Light Cross-Fader

Q1, Q2	SCR (C106B, or similar*)
D1 - D5	1N4003 diode (or similar)
C1	0.5 μF 250 volt capacitor
R1, R3	4.7K $^1/_2$ watt resistor
R2	47K $^1/_2$ watt resistor
F1	Fuse to suit load

*—see text

this circuit. Using the components called for in Table 10-7, each load may be up to 150 watts.

PROJECT 22: SCR POWER CONTROLLER

Figure 10-8 shows a simple power controller circuit built around an SCR. A neon lamp is used to trigger the SCR. Unlike the demonstration circuits presented back in chapter 2, this is a complete, functional circuit.

While there is considerable leeway in the choice of component values, a fairly typical parts list for this project is given in Table 10-8. If you use a different SCR than the one mentioned in the parts list, make

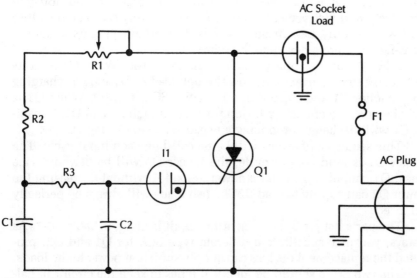

Fig. 10-8 *SCR Power Controller*

Table 10-8   Parts List for SCR Power Controller

Q1	SCR (C106B, or similar*)
I1	neon lamp (NE-2, or similar)
C1, C2	0.068 μF 250 volt capacitor
R1	50K potentiometer
R2	10K $^1/_2$ watt resistor
R3	15K $^1/_4$ watt resistor
F1	Fuse to suit load

*—see text

sure it can handle enough power to safely drive the intended load for the project.

Potentiometer R1 is used in this circuit to control the lighting level, or power output. SCR Q1 is triggered during the positive half-cycles via the neon lamp I1.

In operation, adjust potentiometer R1 until the neon lamp lights, indicating it is conducting. At this point, the lamp will typically be at about half-brilliance. Once you have found this point, reduce the potentiometer's setting, until the load device is receiving the desired amount of power.

Select a fuse to suit the desired load's power requirements. For safety's sake, do not ever omit the fuse from this, or from any other thyristor circuit. Do not use too large a fuse, or it won't serve its intended purpose of protecting the circuit and the operator. Refer to chapter 8.

Remember, this is an ac powered circuit. Use all the necessary precautions. The entire circuit should be well insulated. It should not be possible for the operator to ever accidentally touch any live conductor anywhere in the circuit. Make absolutely sure that no one can possibly receive a shock from this circuit under any imaginable circumstance. No conductor anywhere in any ac power circuit should be exposed.

No amount of possible cost cutting can possibly be worth risking a potentially serious injury, or maybe even a fatality. Don't take foolish and unnecessary chances.

PROJECT 23: AUTOMATIC LIGHT-BALANCE CONTROLLER

This project is a rather unusual variation on the basic light dimmer—power controller idea. In this project, the dimmer circuit senses the ambient light level and automatically adjusts itself to compensate for any uncontrolled changes in the lighting. The circuit automatically balances the lighting, or other load, to a desired level.

Personally, I can't really think of any practical application for this circuit with a load other than electrical lights. However, you might come up with some alternative applications on your own.

Used outside, this circuit will automatically turn on a light at nightfall, and turn it back off at dawn. The schematic diagram for this project is illustrated in Fig. 10-9, with the parts list given as Table 10-9.

Potentiometer R1 serves as the sensitivity control for the circuit. This control sets the desired lighting level when triac Q1 is switched ON.

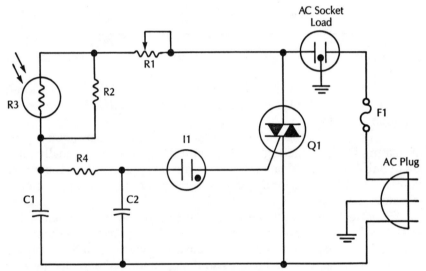

Fig. 10-9 *Automatic Light Balance Controller*

Table 10-9 Parts List for Automatic Light Balance Controller

Q1	Triac (40502, or similar*)
I1	neon lamp (NE-2, or similar)
C1, C2	0.68 250 volt capacitor
R1	10K potentiometer
R2	470K 1/2 watt resistor
R3	photoresistor
R4	15K 1/2 watt resistor
F1	Fuse to suit load

*—see text

A photoresistor, R3, is used to sense the ambient lighting level. Be sure to mount it so it will be exposed to the desired controlling light source, but protected from other, stray light sources, like the load lighting itself, or frequent and uncontrolled momentary shadows.

If used outdoors, make sure that all electrical connections are fully weather-proofed. Extra insulation is always very strongly advisable in any ac circuit used outdoors.

PROJECT 24: LIGHT-OPERATED RELAY

The circuit shown in Fig. 10-10 is another project that is controlled by the ambient lighting level. A typical parts list for this project is given in Table 10-10. A latching relay (K1) is used in this project for on/off control.

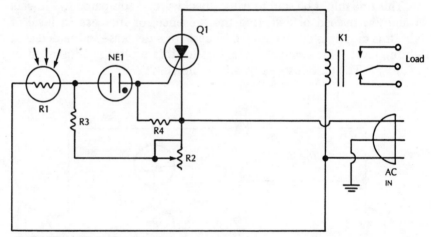

Fig. 10-10   *Light Operated Relay*

Table 10-10   Parts List for Light Operated Relay

Q1	SCR 200 volt, 4 amp, or to suit load
D1	diode (400 PIV)
NE1	neon lamp (NE-83, or similar)
K1	110 VAC latching relay contacts to suit desired load
R1	photoresistor
R2	1 megohm potentiometer
R3	22K $1/2$ watt resistor
R4	100 ohm $1/2$ watt resistor

The photoresistor, R1, is the light sensor for the circuit. If the photoresistor is normally kept dark, then is struck by a beam of light, the SCR, Q1, will be triggered, activating the relay. Being a latching relay, it will stay activated until another trigger pulse or burst of light is detected by the photoresistor.

For obvious reasons, this circuit will work best in a relatively darkened area. It can be remotely controlled with a simple flashlight.

In moderately bright environments, use an infrared sensitive photoresistor for R1. It can be controlled by a hand-held infrared generator, like a wireless remote control for a TV set or VCR.

PROJECT 25: LIGHT INTERRUPTION DETECTOR

Since we've already looked at a couple of light-sensitive thyristor circuits, let's present one more.

This one might be said to work "backwards" compared to Projects 23 and 24. Instead of detecting the presence, or increase in level of light, this circuit is triggered by darkness, or a decrease in the detected lighting level. The schematic diagram for this project is illustrated in Fig. 10-11. The parts list is given in Table 10-11.

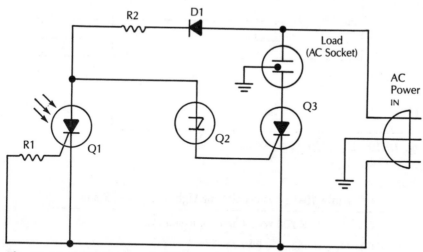

Fig. 10-11   *Light Interruption Detector*

Instead of a simple photoresistor, a Light Activated SCR (LASCR), Q1, is used as the detector in this circuit. The LASCR will be discussed in detail in chapter 12.

For now, all you really need to know is that an LASCR is a specialized SCR which can be gated, or triggered, by an external light source

Table 10-11   Parts List for Light Interruption Detector

Q1	LASCR*
Q2	SUS (2N4490, or similar*)
Q3	SCR to suit load device
D1	diode (1N5059, or similar)
R1	100K $1/2$ watt resistor
R2	18K $1/2$ watt resistor

*—see text

exceeding a specific level of intensity. This project also uses a Silicon Unilateral Switch (SUS), Q2. This component will also be covered in chapter 12.

Ordinarily, the LASCR should be exposed to light. If the light falling on this sensor should be blocked or otherwise removed, the LASCR will fire, triggering SUS Q2, which turns ON the main SCR, Q3. This SCR drives the load device.

When the light beam shining on the LASCR is resumed, the circuit will shut down, turning OFF the load device. The circuit will be activated as long as the LASCR is kept dark.

PROJECT 26: EMERGENCY LIGHT

Power outages, which are all too frequent in many urban areas, can be annoying, or they can be dangerous. There could be problems if certain types of equipment are permitted to go off. Under some circumstances, if the lights go out, someone could get hurt trying to move around in the dark.

One solution is to install an automated emergency light, like the circuit illustrated in Fig. 10-12. If the ac power source is interrupted for any reason, the SCR, Q1, will turn on a small lamp operated from a 12-volt battery. The emergency light will come on almost instantly when there is an ac power failure.

A typical parts list for this project appears in Table 10-12. The 12-volt battery, B1, should be of the rechargeable type. You could use NiCad cells (ten in series), or an automotive storage battery.

When ac power is present, the battery will be recharged through diode D2 and resistor R2. The value of the resistor should be selected to provide the necessary recharging current to the battery. The calculations in this case are simple enough. It's just a matter of using Ohm's Law:

$$R = E/I$$
$$R = 12/I$$

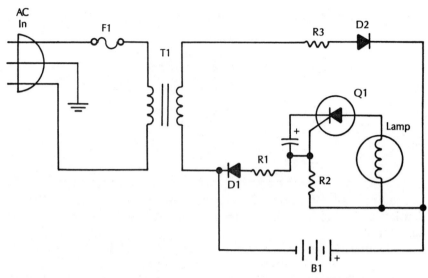

Fig. 10-12 *Emergency Light*

Table 10-12 Parts List for Emergency Light

Q1	SCR (C106Y, or similar*)
D1	diode (GE A14F, or similar)
D2	diode (GE A40F, or similar)
T1	transformer: secondary = 12.6 volts
C1	100 μF 25 volt electrolytic capacitor
R1	100 ohm $^1/_2$ watt resistor
R2	1K $^1/_2$ watt resistor
R3	current limiting resistor to suit battery*
B1	12 volt battery (rechargable)*
F1	Fuse to suit load

*—see text

As long as ac power is present, capacitor C1 will be kept charged through resistor R1 and diode D1. This places a negative voltage on the gate terminal of SCR Q1. Because of this negative gate voltage, the SCR is held in its OFF state, and its load, the lamp, stays off. Under this circumstance, the only thing the circuit does is feed a small trickle charge current to battery B1 through R3 and D2.

Nothing much will happen in this circuit until the ac power source is interrupted for any reason. Diode D1 and resistor R1 are then no longer able to keep the capacitor charged. The capacitor discharges, removing its negative holding voltage from the gate terminal of SCR Q1. The battery voltage from B1 is fed to the gate through resistor R2. This triggers the SCR ON. While the SCR is conducting, current can flow through the lamp, so it lights up, illuminating the area.

Don't expect too much light from this little project. A 12-volt battery won't supply very much power. But it is considerably better than total darkness. In a large area, several separate emergency light circuits are advisable.

The emergency light will stay lit until the battery runs down or until ac power is restored. The circuit automatically shuts itself down when it is no longer needed. The battery can supply power long enough, especially if a large storage battery is used, that power will probably be restored before the emergency light goes out.

The light will certainly last long enough to give you time to find and replace a blown fuse, or to reset a popped circuit breaker. If the power failure is a problem at the electric company, the ac power may be off for a longer time.

If and when ac power is restored, the circuit will automatically reset itself back to its holding mode. The peak ac line voltage reverse-biases the SCR and turns it OFF. By this time, capacitor C1 has charged back up through resistor R1 and diode D1. This charged capacitor places a negative voltage on the SCR's gate that is larger than the positive battery voltage through resistor R2, so the SCR cannot be triggered. It stays OFF until the next time ac power is interrupted.

PROJECT 27: DELAYED OFF LIGHT

Do you have a problem with the kids neglecting to turn off the bathroom light? Or perhaps it's dark in your garage when you park at night. You don't really want to leave the headlights on until you can turn on the garage light, or find your way to the door. It's all too easy to forget to turn the headlights back off and run down the car's battery.

These problems and many similar ones can be solved with this project. Figure 10-13 shows the schematic diagram for a delayed off light-control circuit. A typical parts list for this project is given in Table 10-13.

Switch S1 is an SPDT momentary-action switch. Usually the contacts are in the Normally Closed (NC) position shown in the diagram. When the pushbutton is depressed, the NC connection is broken and the Normally Open (NO) connection is made. When the switch is released, the contacts will go back to their NC position.

Briefly depressing switch S1 will trigger the circuit. Power will be applied to the load for a specific amount of time, determined by the

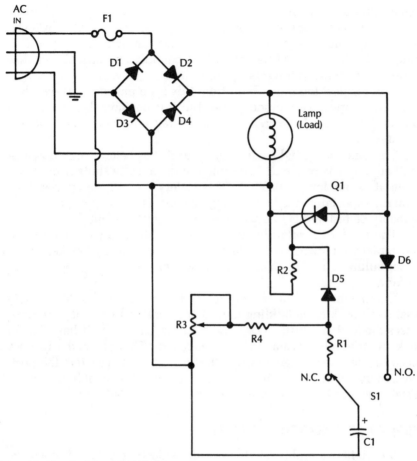

Fig. 10-13 *Delayed Off Light*

Table 10-13   Parts List for Delayed Off Light

Q1	SCR to suit load*
D1 - D6	silicon diodes (1N4004, or similar)
R1	220K ¹/₂ watt resistor
R2	4.7K ¹/₂ watt resistor
R3	1 Megohm potentiometer
R4	2.2K ¹/₂ watt resistor
F1	Fuse to suit load

*—see text

time constant of capacitor C1 and resistors R3 and R4. After this time constant period has passed, the circuit will automatically switch the load back OFF.

R3 is a potentiometer to allow manual control over the load's ON time. Using the component values given in the parts list, the load ON time can be adjusted from a low of a few seconds up to about three minutes.

To increase the maximum time value, increase the value of capacitor C1. Similarly, reducing the value of this component will give the circuit a comparable range of shorter time constants. SCR Q1 for this project, as always, should be selected to suit the desired load's power requirements.

PROJECT 28: ELECTRONIC CROWBAR CIRCUIT

The electronic crowbar circuit shown in Fig. 10-14 is suitable for both ac and dc operation. A typical parts list for this project is given in Table 10-14.

An electronic crowbar is generally used to protect a delicate or vital circuit. In the event of a potentially harmful circuit condition, the electronic crowbar very quickly forms a deliberate short circuit across the power supply lines, causing the system fuse to blow or the circuit breaker to pop, cutting off all power to the circuit.

Why would anyone want to bother with an electronic crowbar? Won't the fuse or circuit breaker protectively open up the circuit anyway?

Well, maybe it will and maybe it won't. It depends on the nature of the original fault. Sometimes there will be a fault condition which may damage certain expensive components while not disturbing the fuse or circuit breaker.

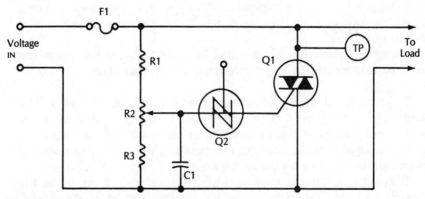

Fig. 10-14 *Electronic Crowbar*

Q1	Triac (to suit load)*
Q2	SBS*
C1	0.1 μF capacitor
R1	10K $^1/_2$ watt resistor
R2	1K potentiometer
R3	1K $^1/_2$ watt resistor
F1	Fuse to suit load

Table 10-14 Parts List for Electronic Crowbar

*—see text

It also takes time for the fuse or circuit breaker to respond to a circuit fault. Even a fast-blow fuse will take a finite amount of time to blow out. The finite time required for the fuse to blow out may be a problem if the overcurrent condition is not too severe.

The overcurrent in this case is high enough to do some damage to the circuit's components, but it is not high enough to generate sufficient heat across the fuse element to cause it to vaporize quickly. In the meantime, the fuse continues conducting the overcurrent to the rest of the circuit. Some delicate semiconductor components could be damaged in that fraction of a second it takes the fuse or circuit breaker to open up. An electronic crowbar circuit, being all electronic, can respond much faster than a mechanical fuse or circuit breaker.

Triac Q1 should be selected to suit the power requirements of the load circuit or device you want to protect with the electronic crowbar. Q2 is a Silicon Bilateral Switch, or SBS. This type of component will be discussed in chapter 12.

The resistance values in the circuit, R_1, R_2, and R_3, determine the operating range for the electronic crowbar. Potentiometer R2 is adjusted to set the precise desired trip value. It might be a good idea to use a trimpot or hidden potentiometer for this control to avoid accidental readjustment.

Using the resistance values listed in Table 10-14, the ac operating range for this electronic crowbar circuit is between about 42 and 84 volts.

The circuit will also work with dc, but the operating range will be altered. Using the same component values, the dc operating range for the electronic crowbar circuit is between about 60 and 120 volts. Of course, the triac must be rated for an operating voltage higher than the upper limit of the circuit's operating range.

To test the electronic crowbar circuit, open up the circuit at the point marked TP (Test Point) in the schematic diagram of Fig. 10-14.

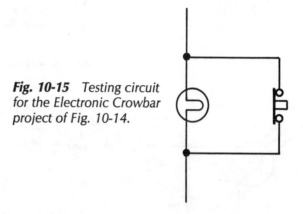

Fig. 10-15 *Testing circuit for the Electronic Crowbar project of Fig. 10-14.*

The test point is by the MT1 terminal of triac Q1. Insert the testing circuit shown in Fig. 10-15, at point TP in Fig. 10-14.

The lamp is a low-power incandescent lamp with a voltage rating that is the same as the circuit's supply voltage. The test switch is a NC SPST pushbutton. Push the button to open the switch and adjust potentiometer R2 to light the lamp at the desired point.

The test circuit (the lamp and the NC push-switch) may be left permanently in the circuit if you choose. As long as the test switch's contacts are closed, circuit operation will not be affected.

While this electronic crowbar circuit offers considerable protection to the load circuit or device, it is not perfect or 100 percent foolproof. Nothing is ever truly perfect, and that definitely includes electronic circuits. The electronic crowbar may not always be triggered, on some very short, infrequent power line transients. Fortunately, such short, infrequent transients generally won't do much permanent harm, although some circuits, especially digital computer circuits, may experience glitches in their operation due to short transients on the ac power lines. For the majority of electronic circuits, this electronic crowbar will provide more than adequate protection.

PROJECT 29: SELF-ACTIVATING NIGHT LIGHT

This project is used to automatically illuminate any area that might go dark. As long as there is sufficient ambient light in the area, the night light will remain off. It will only go on when it is needed, so power is not wasted.

The schematic diagram for this project appears in Fig. 10-16. Table 10-15 shows the parts list. This circuit is entirely straightforward and should be very easy for you to follow.

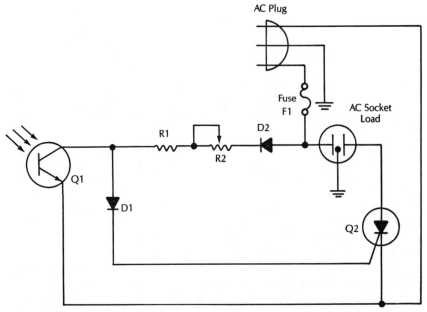

Fig. 10-16 *Self-Activating Night Light*

Q1	NPN phototransistor
Q2	SCR (select to suit load)
D1, D2	diode (1N5059, or similar)
R1	1 Megohm ¹/₂ watt resistor
R2	5 Megohm potentiometer
F1	Fuse to suit load

Table 10-15 Parts List for Self-Activating Night Light

The phototransistor, Q1, serves as the light sensor for the circuit. The requirements for this sensor are quite loose, so almost any npn type phototransistor should work fine. In operation, as long as the phototransistor's sensing surface is sufficiently illuminated, Q1 will shunt SCR, Q2's gate current to ground. The SCR is held in its OFF state, so nothing happens at the load socket.

When the ambient lighting level, as sensed by phototransistor Q1, drops below a specific trigger point however, the phototransistor feeds current into the gate terminal of SCR, Q2, switching it to its ON state. The gate signal is constant as long as the sensor is dark. The SCR will cut OFF on negative half-cycles of the ac power signal, but it will turn back ON at the start of the next positive half-cycle.

When the SCR is turned ON, or triggered, it permits the ac power to flow through the socket, and through whatever load device is plugged into it. Normally, this type of circuit will use some kind of relatively low-power lighting device as the load.

The SCR used in this circuit should be selected to safely supply ample power to the intended load device. Potentiometer R2 sets the sensitivity of the light detector, phototransistor Q1. That is, the resistance of R2 determines how dark the sensor must be to turn ON the SCR and its load. Simply adjust this control for the desired turn-ON point.

PROJECT 30: REMOTE FLASH TRIGGER

SCRs and triacs have certainly been a blessing for the modern photographer. They are ideal for driving flash units, which are used to artificially increase the lighting level of a scene so it can be well photographed. A photographic flash unit produces a burst of bright light while the shutter of the camera is open.

Because the light output is so powerful, we usually don't want it to remain constantly on. Besides being uncomfortable to the eyes, such bright lights can consume quite a bit of power. It's better to turn on the light only when it is needed and then shut it off.

The camera's shutter is open for only a tiny fraction of a second usually, which is all the time we need the extra light to be on.

It is fairly easy to control such bursts of light electronically. Often the lighting unit will be triggered directly by the opening of the camera's shutter. The operation of these devices well deserves the name "flash."

A modern flash unit is usually centered around a xenon tube, or strobe lamp. A xenon tube is a glass tube filled with xenon gas. When the voltage applied to electrodes at opposite ends of the tube exceeds a specific, fairly high value, the xenon gas reacts and emits a brilliant burst of light. An SCR or triac is perfect for firing xenon tubes.

Xenon tubes themselves are often a pain to work with. They are very sensitive to handling and are easily damaged. Even the oil that naturally forms on your fingertips can be harmful. This portion of the flash unit can be purchased ready made for a reasonable cost.

The hobbyist can save quite a bit of money by building his own flash driver circuits, especially those for specialized applications. With all of this in mind, the next three projects are all different types of flash unit drivers for various photographic purposes.

Figure 10-17 shows what may well be the simplest project in this book. There are just two components—a Light Activated SCR (LASCR)—see chapter 12), and a resistor. The parts list, short as it is, is given in Table 10-16.

This circuit is a remote-flash trigger. The main flash is normally triggered directly by the camera. In many photographic sessions, multi-

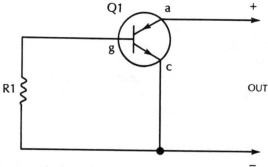

Fig. 10-17 *Remote Flash Trigger*

Table 10-16   Parts List for Remote Flash Trigger

Q1	NPN photo transistor
Q2	SCR (select to suit load)
D1, D2	Diode IN5059
R1	1M ohm $^1/_2$ watt resistor
R2	5M ohm potentiometer
F1	Fuse (select to suit load)

ple lighting units are needed, illuminating the subject from various angles. This project can be used to fire one of these secondary flash units.

No electrical connection is required between this circuit, and the camera, or with the main flash unit. The LASCR is triggered by the burst of light from the main flash unit, firing the secondary flash unit.

Most flash units have polarized synchronization, or "sync" inputs. The anode of the LASCR *must* be connected to the positive lead and the cathode to the negative lead of the flash unit's sync inputs. This circuit will not work if the LASCR is installed backwards.

In most applications, this circuit will be mounted in a small more-or-less light-tight housing, except for a small lens over the sensing area of the LASCR. This lens should be oriented towards the main flash unit so it will receive enough of the light burst to reliably trigger the device.

PROJECT 31: SLAVE FLASH DRIVER

The circuit illustrated in Fig. 10-18 is another approach to driving a secondary, or slave flash unit from the light burst from the main camera controlled flash unit. Again, no electrical connection is required between the slave unit and the camera, or main flash unit.

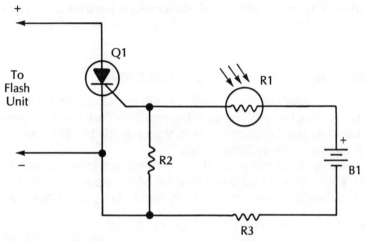

Fig. 10-18 *Slave Flash Driver*

The parts list for this project appears in Table 10-17. Even though this circuit has a higher parts count than Project 30, it is likely to be less expensive. LASCRs tend to be relatively expensive components.

This circuit uses a regular SCR, Q1, (almost any device that can handle enough power to fire the flash unit) and a photoresistor, R1, as the light sensor. Otherwise the principles of operation are basically the same as for Project 30.

Q1	SCR (C106B, or similar*)
B1	1.5 volt battery*
Table 10-17 Parts List for Slave Flash Driver R1	photoresistor
R2	22K ¹/₂ watt resistor
R3	47 ohm ¹/₂ watt resistor

*—see text

Experiment with other values for resistor R2. This resistance determines the overall sensitivity of photosensor R1. The 22K resistor called for in the parts list will be a good choice for most typical photographic applications. In some specialized applications other resistor values may do a better job. Increasing the value of R2 will increase the circuit's sensitivity and vice versa.

Unlike the circuit of Project 30, this circuit requires its own built-in voltage source. It doesn't need much power—just 1.5 volts. You can use an AA or AAA penlight cell. For the smallest possible unit, you could use one of those tiny button-sized batteries used to power digital watches.

PROJECT 32: TIMED MULTIPLE FLASH DRIVER

Some specialized photographic applications call for a stroboscopic effect. This can be achieved by firing multiple flash units in sequence. The basic sequential circuit is shown in Fig. 10-19. The parts list for this project appears as Table 10-18.

The simple output circuit for this project is illustrated in Fig. 10-20. The output circuit is repeated for each output used. Each output circuit drives its own flash unit. Because this part of the circuit is

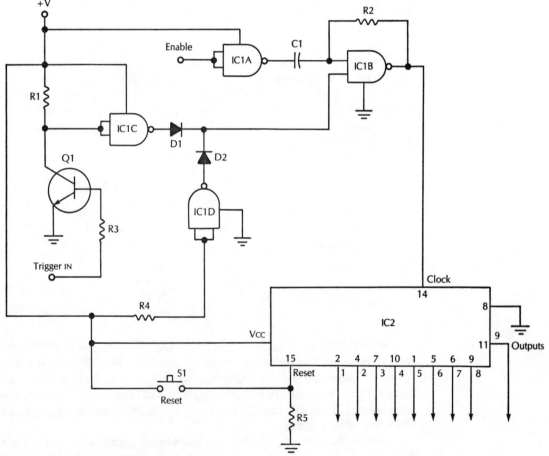

Fig. 10-19 *Timed Multiple Flash Driver*

Table 10-18   Parts List for Timed Multiple Flash Driver

IC1	CD4001 CMOS quad NAND gate
IC2	CD4017 CMOS decade counter
Q1	NPN transistor (2N2222, 2N3904, or similar)
Qx	SCRs (see text)
D1, D2	diode (1N4004, or similar)
R1, R2	1 Megohm $1/4$ watt resistor
R3, R4, R5	100K $1/4$ watt resistor
Rx	1K $1/2$ watt resistors (see text)
S1	Normally Open SPST pushbutton switch

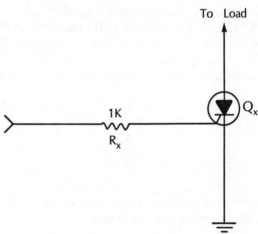

Fig. 10-20 *Output circuit for the Timed Multiple Flash Driver project of Fig. 10-19.*

repeated, specific part numbers are not given in the diagram. The components are identified with an "x" in their identification numbers in the parts list.

The main portion of this circuit is a fairly standard CMOS digital counter circuit. The CD4017 decade counter, IC2, has ten separate and independent outputs. As the counter advances through its count cycle, each output will be activated, firing the associated flash unit individually, in sequence. Only one of the outputs may be active at any given time.

Do not use the "0" output, pin 3, in this application. This output will be activated before the counting process begins. This still leaves you with up to nine sequential outputs to work with:

Output	Pin
1	2
2	4
3	7
4	10
5	1
6	5
7	6
8	9
9	11

I don't know why the original manufacturer of this chip chose such an odd numbering scheme. Nevertheless, it is standardized, now that the CD4017 is widely used in numerous applications.

You do not necessarily need to use all nine of the available outputs. Any unused outputs will be ignored. If you skip one or more outputs you can introduce a delay into the count sequence. For example, if you are using five flash units, you could use straightforward numbering:

1
2
3
4
5

But, if it suits your individual application, you could also skip certain outputs, creating pauses in the sequence:

1
2
(pause)
4
(pause)
(pause)
7
8

Most of the rest of the circuit is pretty straightforward and flexible. None of the component values are particularly critical in this project.

Transistor Q1 can be almost any low-power npn device you happen to have handy. Similarly, almost any silicon diodes can be used for D1 and D2. These components are not at all critical in this circuit.

The SCRs, Qx, used in this project should be selected to provide adequate power to their loads, the flash units. There are no special requirements for these thyristors either.

IC1A and IC2A form a clock, or oscillator to advance the counter. You can alter the frequency, or step rate by changing the value of either capacitor C1 or resistor R2, or both. You could use a potentiometer or trimpot in place of R2, if you so chose.

The circuit will do nothing at all until it receives a suitable trigger pulse on its enable input. This trigger pulse could come from a direct electrical connection to the camera, or main flash unit, or it could be derived from a light sensor to serve as a slave unit.

Once triggered, the counter will step through its entire count cycle, 1 to 9, even if the trigger signal is removed. Only a brief pulse is needed to enable the counter circuit.

Once the counter has completed its sequence, reached a count of 9, it will stop and do nothing, even if the circuit is retriggered. At the end of the count sequence, the circuit must be manually reset by briefly depressing pushbutton S1.

More advanced experimenters may want to modify this circuit for an automated reset function, if this would be appropriate to the specific intended application.

PROJECT 33: TOUCH SWITCH

The circuit shown in Fig. 10-21 can offer a lot of convenience in many practical applications. Any ac powered device can be operated with just the lightest touch of a fingertip to a small touch plate.

The touch plate is simply an exposed piece of conductive material. A small unetched piece of copper-clad PC board is ideal. Touching a fingertip to this touch plate will trigger the circuit. The parts list for this project is given in Table 10-19.

The optoisolators, IC3 and IC4, are used to isolate the dc and ac portions of the circuit. No ac power should ever be able to reach the touch plate under any imaginable circumstances. If the touch plate carried live ac, the shock hazard should be quite obvious. By using optoisolators, there is no electrical connection at all between the two portions of the circuit, so there is no possibility of any component failure creating a dangerous short circuit.

An optoisolator contains an LED light source and a photosensitive device of some type enclosed in a single light-tight housing. The input control circuit drives the light source. The photosensor is part of the output circuit. The only connection between the input and output circuits is a beam of light.

The particular optoisolators used in this project use phototriacs as

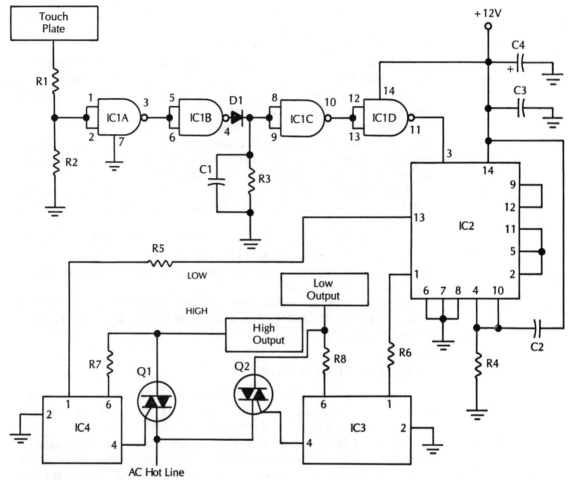

Fig. 10-21   *Touch Switch*

their output devices, so this circuit actually uses four thyristors. You could substitute similar optoisolators for the MOC3010s called for in the parts list. A triac output is necessary for this particular project.

The circuit's output triacs, Q1 and Q2, should be selected to suit the intended load or loads. Notice that there are two outputs for this circuit, labeled LOW and HIGH, respectively. The control circuit around the touch plate is a digital circuit. The touch plate toggles the digital circuitry between its two possible output states—LOW and HIGH.

Only one of the circuit's two outputs is active at any given time. If the LOW output is initially activated, touching the touch plate will turn off the LOW output and turn on the HIGH output. Touching the touch

Table 10-19   Parts List for Touch Switch

IC1	CD4093B CMOS quad Schmitt trigger/NAND gate
IC2	CD4013B CMOS dual D-type flip-flop
IC3, IC4	MOC3010 triac output optocoupler
Q1, Q2	Triac
D1	diode (1N914, or similar)
C1 - C3	0.1 μF capacitor
C4	5000 μF 20 volt electrolytic capacitor
R1, R4	100K $^1/_4$ watt resistor
R2	10 Megohm $^1/_4$ watt resistor
R3	680K $^1/_4$ watt resistor
R5, R6	270 ohm $^1/_2$ watt resistor
R7, R8	180 ohm $^1/_2$ wattt resistor

plate again will turn off the HIGH output and turn the LOW output back on, returning the circuit to its initial condition.

Depending on the specific application, either one or the other or both of the outputs may be used. While the schematic diagram for this project may look a bit complex, you should run into no major problems in building the circuit. You will undoubtedly notice that the parts list for this project is admittedly a little long, but the components are all fairly common and not too expensive. You shouldn't have any major problems gathering together all of the components necessary to construct this circuit.

PROJECT 34: SCR OSCILLATOR

This project is a bit of an oddball, which may even surprise some experienced electronic technicians. Normally SCRs are used in power controller applications. With a little creative circuitry it can also be forced to function as an oscillator. The circuit for the SCR oscillator appears in Fig. 10-22. A typical parts list for this project is given in Table 10-20.

Using the component values called for in the parts list, the output frequency is quite low. LED, D1, is used in this circuit as an output indicator. When the circuit is operating, you will see the LED flash on and off. You are strongly encouraged to experiment with other component values in this project.

If the frequency is made too high, the LED will appear to be continuously lit. Actually, it is still blinking on and off, but it is turning on and off far too rapidly for the human eye to see the individual flashes. Each successive flash will blend into the one before it and the one after

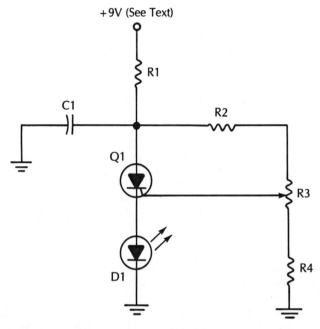

Fig. 10-22 *SCR Oscillator*

Q1	SCR (see text)
D1	LED (see text)
C1	0.1 μF capacitor (see text)
R1	100K resistor
R2, R4	220K resistor
R3	500K potentiometer

Table 10-20 Parts List for SCR Oscillator

it, so the LED will look like it is continually lit. This effect is known as the "persistence of vision effect." This same effect is responsible for the illusion of movement when you are watching a movie.

For higher oscillator frequencies, an LED doesn't make a very good output indication device. You might consider replacing the LED with a small loudspeaker of some kind. The output frequency will then be heard as a tone. However, if the output frequency is too low, down in the LED flashing range, you won't be able to hear anything from the speaker.

Technically this type of circuit is known as a relaxation oscillator. In this type of oscillator, a capacitor is gradually charged up and then rapidly discharged through a resistance. This is repeated over and over,

once for each output cycle. The output of a relaxation oscillator is typically a rather odd semi-pulse waveform, like the one illustrated in Fig.
10-23.

The SCR in this circuit is self-triggered. Actually the thyristor is
being used more as if it was a four-layer diode. You should remember
that a four-layer diode is identical to an SCR without a gate terminal,
and can be used in oscillator circuits.

Almost any SCR can be used in this circuit, although some devices
may require a different supply voltage. If your SCR relaxation oscillator
doesn't work, try experimenting with different supply voltages. Typically, the power supply voltage should be kept between about + 6 volts
and + 15 volts.

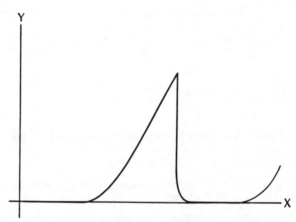

Fig. 10-23 *Typical output waveform for the SCR Oscillator project of Fig.
10-22.*

PROJECT 35: DIGITALLY CONTROLLED AC SWITCH

With this project you can use a computer or a simpler digital signal
source to turn any ac load device on and off. This capability is particularly useful in automation systems.

The circuit for the digitally controlled ac switch is shown in Fig.
10-24. The parts list for this project appears in Table 10-21. As usual,
the triac should be selected to suit the intended load device for the
project.

An optoisolator, IC1, is used to keep the digital circuit and the ac
power circuit entirely separate from one another. Optoisolators were
also employed in Project 23, earlier in this chapter.

The optoisolator used in this project uses an npn phototransistor as
its output device. A TIL112 optoisolator is recommended, but any simi-

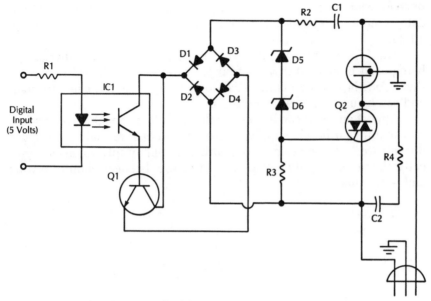

Fig. 10-24   *Digitally Controlled ac Switch*

Table 10-21   Parts List for Digitally Controlled AC Switch

IC1	TIl112 optoisolator*
Q1	NPN transistor (2N3904, or similar)
Q2	Triac*
D1 - D4	diode (1N4002, or similar)
D5, D6	zener diode 5.6 volts*
C1	0.36 µF 500 volt capacitor
C2	0.1 µF 250 volt capacitor
R1	470 ohm 1/4 watt resistor
R2	1K 1 watt resistor
R3	1K 1/2 watt resistor
R4	100 ohm 1/2 watt resistor

*—see text

lar device may be substituted. The requirements for the optoisolator here are not too critical.

Transistor Q1 is also not particularly critical for this circuit. Almost any npn transistor may be substituted for this component.

The impedance of capacitor C1 effectively controls the gate current for triac Q2. On each half-cycle, the triac is ac triggered through this capacitor, resistor R1, and diodes D5 and D6. Notice that these diodes are zener types, with a rating of 5.6 volts. Zener diodes with other voltage ratings may be substituted, but the operation of the circuit will be affected. Do not attempt a substitution for diodes D5 and D6 unless you definitely know exactly what you are doing.

When transistor Q1 is held OFF, by the input voltage from optoisolator, IC1, triac Q2 can be retriggered on each half-cycle. The triac will conduct current through to the load.

When the optoisolator turns transistor Q1 ON, the triac is held OFF, and is not permitted to retrigger. No current reaches the load socket, and the load device is turned OFF.

A logic 0 digital signal turns OFF transistor Q1, and the triac and its load. To turn the transistor ON, and the triac and load OFF, the digital input signal should be a logic 1.

The optoisolator prevents any possibility of the high ac voltage from backing up into the computer or other controller and doing any circuit damage, or creating a shock hazard.

PROJECT 36: TIMED SWITCH

Figure 10-25 shows a rather interesting thyristor circuit. It is an electronic switch circuit which can be activated for a preset period of time. A typical parts list for this project is given in Table 10-22.

When switch S1 is momentarily depressed closed, the load device will be turned ON. Releasing switch S1 will have no effect on the load device. The load will remain ON for a specific period of time, then it will automatically be switched back OFF.

The timing period in this circuit is controlled by a 555 timer, IC2. This is a very popular and versatile IC which is quite easy to use. The

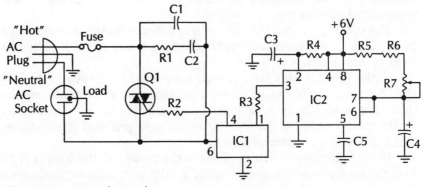

Fig. 10-25 *Timed Switch*

Table 10-22 Parts List for Timed Switch

IC1	Optoisolator—Triac output (MOC3010, or similar*)
IC2	555 timer
Q1	Triac*
C1, C5	0.01 μF capacitor
C2	0.1 μF capacitor
C3	5 μF 50 volt electrolytic capacitor
C4	3.3 μF 50 volt electrolytic capacitor
R1	47 ohm $^1/_2$ watt resistor
R2, R3	390 ohm $^1/_2$ watt resistor
R4	10K $^1/_4$ watt resistor
R5, R6	10 Megohm $^1/_4$ watt resistor
R7	1 Megohm potentiometer

*—see text

time constant for the 555 is:

$$T = 1.1\ RC$$

In this particular circuit, C is capacitor C6. R is the series combination of resistors R5, and R6 and potentiometer R7. The potentiometer allows the user to manually adjust the ON-time. In this project it would almost certainly be a good idea to make up a calibrated dial-plate for potentiometer R7. This dial-plate can be made by trial and error with a stopwatch. Simply time how long the load device remains ON at each setting of the potentiometer.

An optoisolator, IC1, is used to keep the dc and ac portions of the circuit entirely separate. This protects the user from a possible shock hazard, and IC2 from potential damage. This optoisolator has a triac as its output device. The MOC3010 is pretty widely available, but if you run into problems obtaining this unit, you could substitute a similar triac-output optoisolator.

The triac, Q1, should of course be selected with the intended load in mind. Any triac that can supply sufficient power to the load can be used in this circuit.

Notice that IC2 calls for a dc supply voltage of + 6 volts in this circuit. Actually, any dc voltage source from about + 5 to + 12 volts may be used. The exact supply voltage used may possibly have a minor effect on the timing period of the circuit, but in most practical applications, this will not be a significant effect.

For the component values given in the parts list, the load will be held ON for a period of approximately a minute or two, depending on the actual setting of potentiometer R7.

PROJECT 37: AUTOMATIC FADING LIGHT DIMMER

A light suddenly flashing on or off can be annoying, especially if it is the main lighting source in a darkened area. A light dimmer can be used, but can be a nuisance to set to the desired level. You may not want or need the ability to vary the light's overall brightness.

The circuit shown in Fig. 10-26 is a rather unusual variation on the basic light dimmer. When the lights are turned on via SPDT switch S1, they gradually fade up from full-off to full-on, rather than snapping on immediately.

Similarly, when switch S1 is moved to its off position, the lights will slowly fade back to full-off. In between, the light will be at full brilliance. A suitable parts list for this unusual thyristor project is given in Table 10-23.

The control timer is made up of transistors Q1, Q2, and Q3, and their associated components. Notice that Q3 is a Unijunction Transistor, (UJT), which is a different type of thyristor. UJTs will be discussed in chapter 11.

The timing section is separated from the ac controller section in this circuit by optoisolator IC1. Once again, we are using the MOC3010

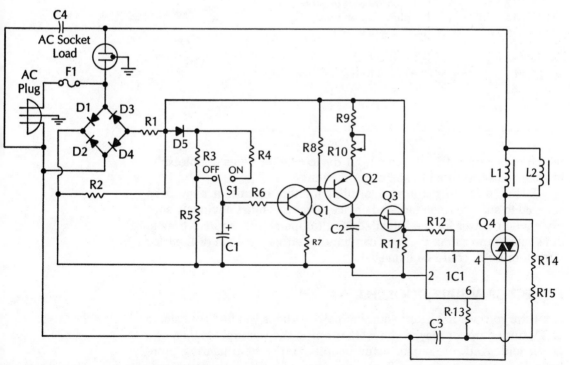

Fig. 10-26   *Automatic Fading Light Dimmer*

Table 10-23   Parts List for Automatic Fading Light Dimmer

IC1	Optoisolator—triac output (MOC3010, or similar*)
Q1	NPN transistor (2N3904, or similar)
Q2	PNP transistor (2N3906, or similar)
Q3	UJT transistor (2N4871, or similar)
Q4	Triac*
D1 - D4	diode (1N4004, or similar)
D5	1N4004
L1, L2	100 μH RF choke
C1	100 μF 50 volt electrolytic capacitor
C2, C3, C4	0.1 μF capacitor
R1	10K $^1/_4$ watt resistor
R2	3.3K $^1/_4$ watt resistor
R3	1.2 Megohm $^1/_4$ watt resistor
R4, R5	150K $^1/_4$ watt resistor
R6	220K $^1/_4$ watt resistor
R7	56K $^1/_4$ watt resistor
R8	10K $^1/_4$ watt resistor
R9	1K $^1/_4$ watt resistor
R10	5K potentiometer
R11	220 ohm, $^1/_4$ watt resistor
R12	100 ohm $^1/_4$ watt resistor
R13, R15	180 ohm $^1/_2$ watt resistor
R14	2.2K $^1/_2$ watt resistor
S1	SPDT switch
F1	Fuse to suit load

*—see text

which uses a phototriac as its output device. Other types of optoisolators may be substituted in some applications.

Triac Q4 is the main ac controller for this circuit. It should be selected to suit the intended load device. The triac must obviously supply enough power for the intended load to operate. If the load tries to draw more power than the triac can handle, either the load or the triac circuit, or both, could be damaged.

PROJECT 38: VISIBLE DOORBELL

Have you ever not heard your doorbell? Perhaps you had your stereo or TV turned up, or maybe you were running the vacuum cleaner, or a power tool. And, of course, many people simply have hearing problems. In any case, the sound from a doorbell can sometimes be missed.

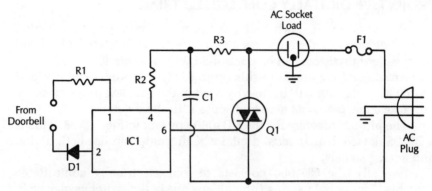

Fig. 10-27 *Visible Doorbell*

The circuit shown in Fig. 10-27 will turn on the light, or other electrical device when the doorbell is rung. Even if you can't hear it, you can see it. A suitable parts list for this useful project appears as Table 10-24.

An optoisolator with a phototriac output, IC1, eliminates any direct electrical connection between the doorbell circuit and the ac power controller circuit. This maximizes safe operation of the project, and minimizes the chances for a dangerous shock hazard.

Of course, triac Q1 should be selected to provide adequate power to the intended load device. Since the output load is a small lamp in this application, an inexpensive, low-power thyristor may be used in this circuit.

Table 10-24   Parts List for Visible Doorbell

IC1	Optoisolator—triac output (MOC3010, or similar*)
Q1	Triac*
D1	diode (1N4002, or similar)
C1	0.25 µF capacitor
R1	390 ohm $^1/_2$ watt resistor
R2	220 ohm $^1/_2$ watt resistor
R3	1.2K $^1/_2$ watt resistor
F1	Fuse to suit load

*—see text

PROJECT 39: DIGITALLY CONTROLLED TRIAC

This project is somewhat similar in concept to Project 35. A digital control signal is used to turn ON a thyristor. Up to four individual and independent thyristors can be controlled with a single IC.

An important element in this project is the CD4066 quad bilateral switch IC. This chip will be heavily used in the next few projects, so we will take a few moments here to discuss it in some detail.

The pin-out diagram for the CD4066 appears in Fig. 10-28. This is a CMOS device. Unlike most digital ICs, this chip can use both digital and analog signals.

Basically, the CD4066 consists of four electrically controllable switches. The signal passing through any one of the switches may be in either digital or analog form, just as with an ordinary mechanical switch. The switch is not polarized. Signals may pass through it in either direction. This is why the term *bilateral* is used in the name of this device. *Bilateral* simply means two-sided.

Some technical literature refers to the CD4066 as a "quad analog switch." There is no difference except in the name used.

You may come across a similar device with the type number CD4016, especially on the surplus market. The CD4066 and the

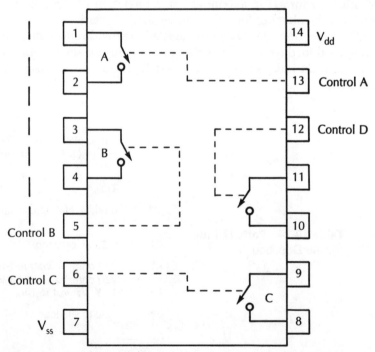

Fig. 10-28 *The CD4066 permits digital control over analog signals.*

CD4016 are very similar. Essentially, the CD4066 is the "new and improved" version of the CD4016. The two chips are pin for pin compatible. The CD4016 may usually be substituted for the CD4066 with no other changes to the circuit. The chief difference between the two chips is that the CD4066's ON resistance when an internal switch is closed is significantly lower than for the CD4016.

Each switch is individually controllable via a single bit digital signal. A logic 0 (LOW) on a given switch's control line will cause that switch to be open (OFF). Similarly, a logic 1, HIGH, on the control line will cause the appropriate switch to close, ON.

Each of the four internal switches and their associated control inputs are summarized as follows:

Switch	In/Out Pins		Control Input
A	1	2	13
B	3	4	5
C	8	9	6
D	10	11	12

Like most CMOS chips, the CD4066 can be operated from a fairly wide range of power supply voltages. Anything from about + 3 volts to + 15 volts will do. For the best and most reliable results, a well-regulated supply voltage is recommended. As a rule, higher power supply voltages will offer better overall circuit performance. I find + 9 volts to + 12 volts generally to be the best choice.

The control signal must be digital pulse, as recognized by CMOS devices. The requirements are, again, pretty flexible. The LOW signal should be a little above ground potential, while the HIGH signal should be a bit under the circuit's supply voltage.

The signal applied across any of the switches may be almost anything. Either analog or digital signals may be switched by the CD4066. There are some important restrictions on the signal to be switched, however. The frequency response is limited to under about 40 MHz. This should not be a problem in most applications, except in very high speed computer systems.

A more important restriction is the maximum voltage which can be safely applied across the CD4066's switches. There is an absolute maximum rating of + 7.5 volts peak across any one switch. This is a *peak*, not an *rms* value. It must never be exceeded at any time, or the chip will almost certainly be damaged.

The 7.5 volt rating is an absolute maximum. To be on the safe side, it would be a very good idea not to let the voltage across any of the switches ever exceed half of the power supply voltage. For example, if you are powering the CD4066 from a 9-volt supply, do not attempt to switch more than 4.5 volts.

The voltage on any of the pins of a CMOS IC, like the CD4066, should never exceed the power supply voltage. Do not try to switch 7.5 volts if you are powering the chip from a 3-volt supply. You'll just burn out the CD4066.

With the CD4066, digital control of thyristors is very easy. Just place the digitally controlled switch in the thyristor circuit's trigger line. When the switch is open, no trigger signal can reach the thyristor, so it remains in its OFF state. Closing the switch permits trigger signals to reach the thyristor, and the circuit performs as if it was an ordinary unswitched design.

Since the CD4066 contains four independent switches on a single chip, you can use one IC to digitally control up to four separate thyristors, as illustrated in Fig. 10-29.

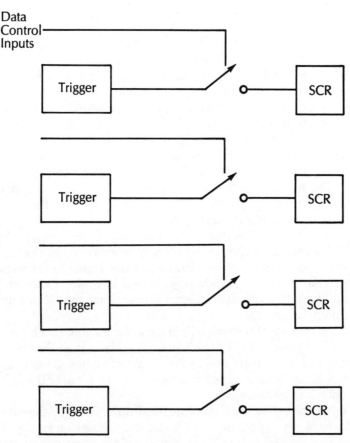

Fig. 10-29 *A CD4066 can control up to four thyristor circuits.*

A practical digitally controlled thyristor circuit is shown in Fig. 10-30. In this circuit, the controlled thyristor is a triac. Much of this circuit should look quite familiar to you. It is basically a modification of Project 19, from earlier in this chapter.

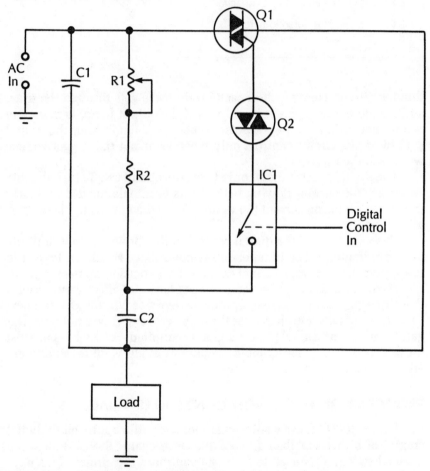

Fig. 10-30 *Digitally Controlled Triac*

The basic circuit here is a light dimmer. The digital control function applies only to when the triac is permitted to turn on its load light. The light will be on as long as the digitally controlled switch is held closed by a logic 1 signal on the appropriate control input pin. When the switch is open, a logic 0, or LOW control input, the triac is disabled, and the load light stays off.

The triac is retriggered on each successive input power cycle as long as the control switch is held closed, just as in an ordinary light

IC1	CD4066 quad bilateral switch	
Q1	Triac to suit load	**Table 10-25 Parts List for Digitally Controlled Triac**
Q2	Diac to match Triac (Q1)	
C1, C2	0.1 μF 250 volt capacitor	
R1	250K potentiometer	
R2	3.3K ½ watt resistor	

dimmer circuit. How much of each power cycle gets through the triac, and therefore, the apparent brightness of the output lamp, is manually determined by the setting of potentiometer R1. Remember, the digital control of this circuit controls only whether or not the trigger signals will reach the thyristor.

The digital control signal only turns the circuit ON. To turn the thyristor OFF, the current flowing between its terminals must drop below the device's holding current (I_h) value. This is the normal mode of operation for a thyristor.

The advantage of the digital control in this project is that multiple dimmer circuits can be independently controlled. Feeding a logic 1 to the appropriate control terminal enables the associated thyristor circuit.

Feeding a logic 0 to the same control terminal effectively disables that thyristor circuit's trigger input. The thyristor will cut OFF the next time its input power cycle passes through zero, and it will not be retriggered until a new digital logic 1 signal is supplied. A suitable parts list for this project is given in Table 10-25. You might want to experiment with alternative component values.

PROJECT 40: DIGITAL ON/OFF CONTROLLED TRIAC

Figure 10-31 shows a somewhat more versatile approach to digital control of a thyristor than Project 39. Once again, the CD4066 quad bilateral switch IC is used to facilitate control of the circuit. This unusual chip was discussed in some detail in the text for Project 39.

Two digital control inputs are provided for each thyristor circuit in this project. One enables or disables the trigger input, as in the earlier version. The other digitally controlled switch enables or disables the load itself. The load may be turned OFF at any point. There is no need to wait until the next zero crossing of the input power cycle. This function can be very useful with very low-frequency power waveforms, or when the thyristor is operating with dc signals.

Two CD4066 quad bilateral switch ICs are shown here. This permits full on/off control over up to four independent thyristor circuits.

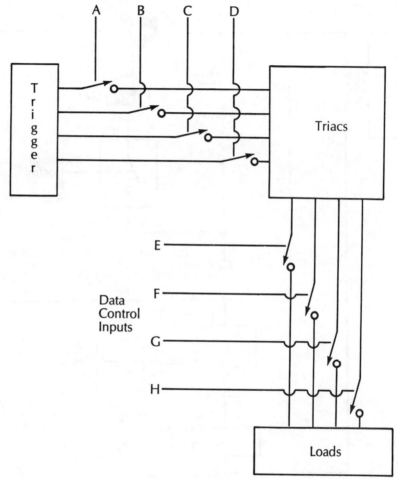

Fig. 10-31 *Two sets of CD4066 switches can provide both* ON *and* OFF
control.

Fig. 10-32 shows a practical schematic diagram for this project. A typical parts list appears as Table 10-26.

Basically, this is the same triac light dimmer circuit we used in
Project 39. The chief difference here is the added control switch, IC2, in
the load line. There are a few other significant differences in this circuit
which you should take notice of.

The first is the use of input transformer T1. In some applications,
an actual transformer may not be needed. It is shown here primarily to
indicate that the full ac house current voltage, 120 volts, must *not* be
applied to this circuit and its load. This circuit is only suitable for use
with low voltages.

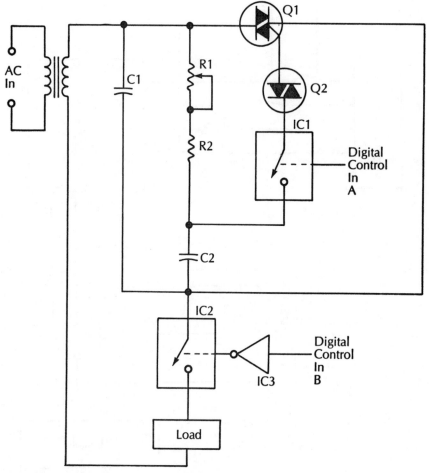

Fig. 10-32 *Digital ON/OFF Controlled Triac*

Table 10-26 Parts List for Digital ON/OFF Controlled Triac

IC1, IC2	CD4066 quad bilateral switch
IC3	CD4049 hex inverter
Q1	Triac to suit load
Q2	Diac to match triac (Q1)
C1, C2	0.1 μF 250 volt capacitor
T1	power transformer—see text
R1	250K potentiometer
R2	3.3K $^1/_2$ watt resistor

You must always be careful never to exceed the maximum voltage which can be safely applied across the CD4066's switches. There is an absolute maximum rating of + 7.5 volts peak across any one switch. This is a *peak*, not an *rms* value. It must never be exceeded at any time, or the chip will almost certainly be damaged.

The 7.5 volt rating is an absolute maximum. To be on the safe side, it would be a very good idea not to let the voltage across any of the switches ever exceed one-half of the power supply voltage.

The voltage on any of the pins of a CMOS IC, like the CD4066, should never exceed the power supply voltage. Do not try to switch 7.5 volts if you are powering the chip from a 3-volt supply. You'll just burn out the CD4066.

Another important point to consider in this circuit is IC3. This is a digital inverter. Its output is always at the opposite state as its input. That is, if the input is LOW, the output will be HIGH, and vice versa.

This component will often be necessary to keep the load control switch, IC2, in a normally closed state. Remember, the digitally controlled switches are closed by a logic 1 and opened by a logic 0 signal. The inverter, IC3, reverses this. Now, a logic 0 control signal, near ground potential, holds the load switch closed. A logic 1 signal opens up this switch and disables the load.

PROJECT 41: DIGITALLY CONTROLLED LIGHT DIMMER

Projects 39 and 40 were digitally controlled light dimmer circuits, but the digital control function could just turn the circuit ON and OFF. The actual dimming function—how much power reaches the load—was not under digital control in these projects.

As you've probably guessed, this project does offer digital control over the dimming function. Let's take a quick refresher look at the basic triac light dimmer circuit, shown again in Fig. 10-33.

The length of time per cycle the thyristor is in its ON state, permitting power to reach the load device, is manually controlled by potentiometer R1. Changing the resistance changes the amount of the input power that is fed through to the load.

To achieve digital control of this circuit, we need some way to digitally control an analog resistance value. It may not seem obvious, but the CD4066 quad bilateral switch IC, which we have been working with in the last couple of projects, will do the trick quite nicely.

The CD4066 will not permit smooth linear control of a resistance value, of course, but digital signals, which we are using as control signals here, are step functions anyway.

The basic CD4066 digitally controlled resistance circuitry is illustrated in Fig. 10-34. It's really quite simple when you look at it. The

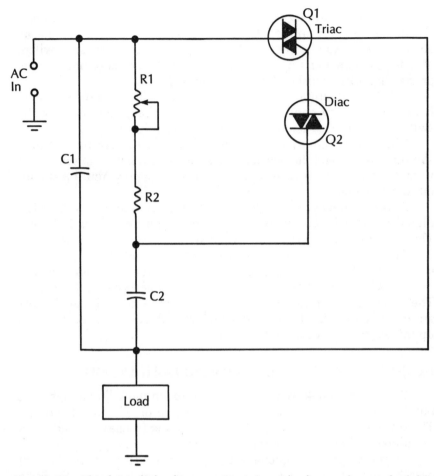

Fig. 10-33 *This basic light dimmer circuit is at the heart of several of the projects in this section.*

CD4066 is used to switch various fixed resistances in and out of the circuit. For example, when switch B is closed, logic 1 control signal, resistor R_b will be part of the active circuit. When switch B is open, logic 0 control signal, resistor R_b will be effectively ignored by the circuit.

Since the CD4066 contains four digitally controlled switches, up to four resistors can be controlled with a single chip.

More than one switch may be closed at once. If this happens, the total effective resistance will be the parallel combination of the selected resistors. With four control bits, we have sixteen possible combinations, ranging from 0000 (no resistors selected) up to 1111 (all four resistors

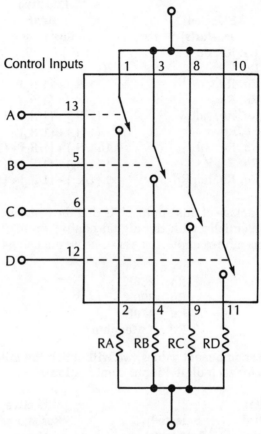

Fig. 10-34 *The CD4066 can be used to set up a digitally controlled resistance.*

selected). Each digital control input results in a unique resistance value:

Digital Control Value	Selected Resistor(s)	Effective Total Resistance
DCBA		
0000	none	∞ (infinity)
0001	Ra	R_a
0010	Rb	R_b
0011	Ra, Rb	$1/[(1/R_a) + (1/R_b)]$
0100	Rc	R_c
0101	Ra, Rc	$1/[(1/R_a) + (1/R_c)]$
0110	Rb, Rc	$1/[(1/R_b) + (1/R_c)]$

Digital Control Value	Selected Resistor(s)	Effective Total Resistance
0111	Ra, Rb, Rc	$1/[(1/R_a) + (1/R_b) + (1/R_c)]$
1000	Rd	R_d
1001	Ra, Rd	$1/(1/R_a) + (1/R_d)$
1010	Rb, Rd	$1/(1/R_b) + (1/R_d)$
1011	Ra, Rb, Rd	$1/(1/R_a) + (1/R_b) + (1/R_d)$
1100	Rc, Rd	$1/(1/R_c) + (1/R_d)$
1101	Ra, Rc, Rd	$1/(1/R_a) + (1/R_c) + (1/R_d)$
1110	Rb, Rc, Rd	$1/(1/R_b) + (1/R_c) + (1/R_d)$
1111	Ra, Rb, Rc, Rd	$1/(1/R_a) + (1/R_b) + (1/R_c) + (1/R_d)$

One disadvantage of this approach is that the resistance value does not change sequentially with the digital control value. This is made clearer with a specific example. Let's assume we are using the following resistor values:

Ra	220K
Rb	390K
Rc	470K
Rd	1 Megohm

Using these component values, we will obtain the following effective resistances for each digital input combination:

Digital Control Value	Selected Resistor(s)	Effective Resistance Value
0000	one	∞ (infinity)
0001	Ra	220K
0010	Rb	390K
0011	Ra, Rb	141K
0100	Rc	470K
0101	Ra, Rc	150K
0110	Rb, Rc	213K
0111	Ra, Rb, Rc	108K
1000	Rd	1000K (1 Megohm)
1001	Ra, Rd	180K
1010	Rb, Rd	281K
1011	Ra, Rb, Rd	123K
1100	Rc, Rd	320K
1101	Ra, Rc, Rd	130K
1110	Rb, Rc, Rd	173K
1111	Ra, Rb, Rc, Rd	98K

Notice that the lowest combined resistance value is obtained when all four resistors are switched into the circuit. Remember, for parallel resistances, the total effective value is always less than for any of the individual component values.

The complete schematic diagram for this digitally controlled light dimmer project is shown in Fig. 10-35. Notice that all we have done here is substitute the CD4066 controlled switched resistor network of Fig. 10-34 in place of the manual potentiometer of the original circuit (Fig. 10-33). A suitable parts list for this project appears in Table 10-27. Feel free to experiment with alternative component values.

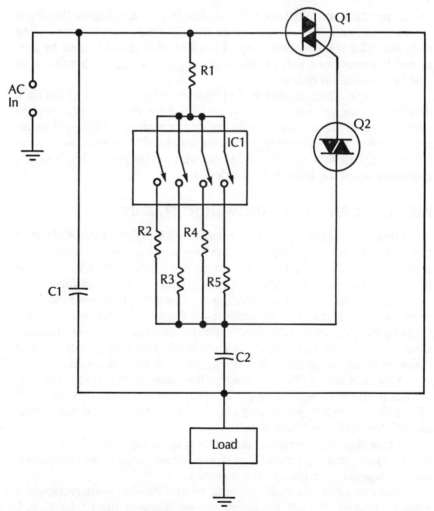

Fig. 10-35 *Digitally Controlled Light Dimmer*

Table 10-27 Parts List for Digitally Controlled Light Dimmer

IC1	CD4066 quad bilateral switch
Q1	Triac to suit load
Q2	Diac to match triac (Q1)
C1, C2	0.1 μF 250 volt capacitor
R1	180K $^1/_2$ watt resistor
R2	220K $^1/_2$ watt resistor
R3	390K $^1/_2$ watt resistor
R4	1 Megohm $^1/_2$ watt resistor

In practical applications for this circuit, you must limit the input power level, and the voltage across the CD4066 digitally controlled switches. The voltage across any of the switches should never be permitted to exceed one-half of the chip's supply voltage. 7.5 volts is an absolute maximum rating.

Either use a transformer to drop the ac voltage applied to the light dimmer circuit, or make sure that resistor R1's value is very large in comparison to the values of resistors R2 through R5. This will cause most of the voltage to be dropped across resistor R1 before it reaches the IC. Triac Q1 in this circuit, of course, should be selected to suit the requirements of the intended load device.

PROJECT 42: PSEUDO-RANDOM LIGHT FLASHER

I think you'll find this next project is a lot of fun. It certainly isn't the most useful and practical thyristor application in this book, although it could come in handy for an eye-catching display, perhaps for advertising purposes. Mostly, though, this circuit is just for fun.

This project is another variation on the basic idea we have been working with in the last few projects—a digitally controlled thyristor. Once again, as in Project 39, four thyristors are trigger-enabled/disabled through a CD4066 quad bilateral switch IC. The trigger functions of these thyristors are more or less randomly enabled and disabled.

A block diagram of this project is illustrated in Fig. 10-36. For convenience, the schematic diagram is split into two parts. Figure 10-37 shows the oscillator circuit, and Fig. 10-38 shows the thyristor control circuit. You will need four of each.

A complete typical parts list for this project appears in Table 10-28. By all means, you are encouraged to experiment with other component values, especially in the oscillator sections

This circuit is not at all difficult to understand. Four rectangular wave oscillators, IC1 – IC4, control the switches of the CD4066, IC5. When a rectangular wave signal goes LOW, its switch is open, and when

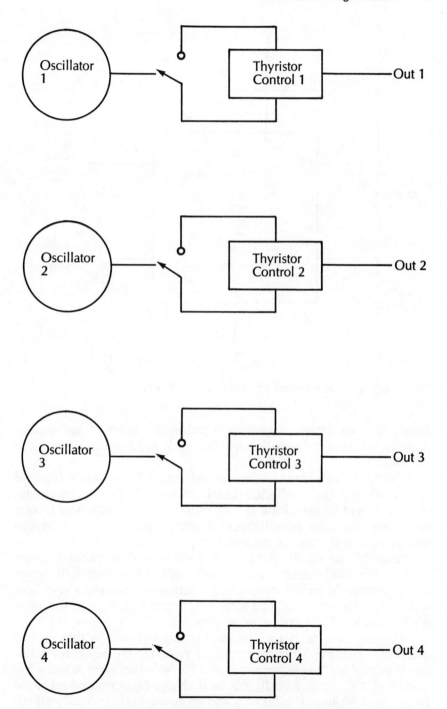

Fig. 10-36   *Pseudo-Random Light Flasher—Block diagram*

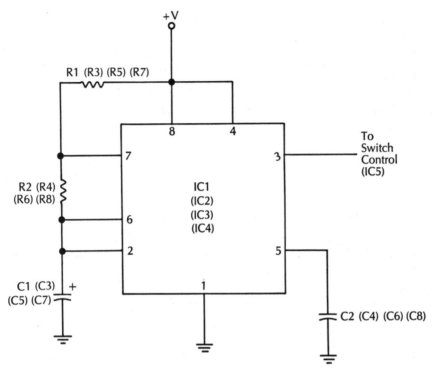

Fig. 10-37 *Pseudo-Random Light Flasher-Oscillator circuit*

the signal goes HIGH, the associated switch is closed. A rectangular wave, of course, continuously switches back and forth between these two states.

The most likely loads for this project would be low-power lights of some kind, perhaps of differing colors. Because the four oscillators are set for different frequencies and duty cycles—ratio of HIGH time to total cycle time—the four output lights will turn ON and OFF in a very irregular, and seemingly random pattern.

Actually, the circuit produces a pseudo-random pattern, rather than a truly randomized output. Eventually the pattern will repeat itself. Depending on the relationships between the various oscillator signals, the sequence might be quite long before it repeats, so the effect is almost as good as a truly random pattern.

The only real restriction on the oscillator signals is that no two oscillators should be set at harmonically related frequencies. For example, if oscillator C has a frequency of 2 Hz, and oscillator A has a frequency of 4 Hz, then load light A will always blink on and off twice every time load light C blinks on and off once. This obviously repetitious pattern would defeat the point of this project.

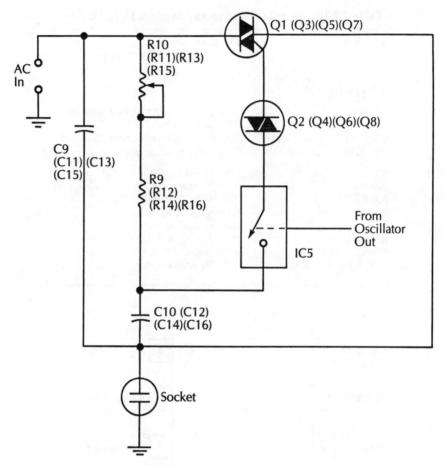

Fig. 10-38   *Pseudo-Random Light Flasher—Thyristor control circuit*

Each of the oscillator circuits is a simple astable multivibrator built around a 555 timer. You could reduce the parts count of the whole project somewhat by using two 556 dual timers in place of the four 555 timer ICs called for in the parts list. A 556 contains two complete 555-type timers in a single package. The two chips are functionally identical. The only thing to watch out for in such a substitution is in using the appropriate pin numbers. To help you in this, the pin-out diagram of a 555 timer IC is shown in Fig. 10-39, and the pin-out diagram of a 556 dual timer is given in Fig. 10-40.

Take another look at the oscillator circuit of Fig. 10-37. There are three frequency determining components—capacitor C1, and resistors R1 and R2. The values of these components also determine the duty cycle of the rectangular waveform.

Table 10-28 Parts List for Pseudo-Random Light Flasher

IC1 - IC4	555 timer (see text)
IC5	CD4066 quad bilateral switch
Q1, Q3, Q5, Q7	Triac to suit load
Q2, Q4, Q6, Q8	Diac
C1, C3	50 μF 35 volt electrolytic capacitor
C2, C4, C6, C8	0.01 μF capacitor
C5, C7	100 μF 35 volt electrolytic capacitor
C9 - C16	0.1 μF 250 volt capacitor
R1	12K $^1/_4$ watt resistor
R2, R4	3.3K $^1/_4$ watt resistor
R3, R5	22K $^1/_4$ watt resistor
R6	4.7K $^1/_4$ watt resistor
R7	33K $^1/_4$ watt resistor
R8	10K $^1/_4$ watt resistor
R9, R10, R12, R14	250K potentiometer
R11, R13, R15, R16	3.3K $^1/_2$ watt resistor

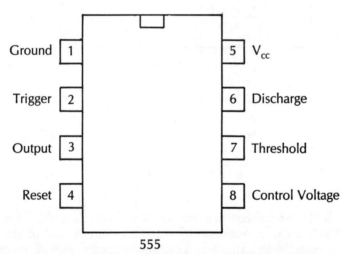

Fig. 10-39 *The 555 timer IC is used in the oscillator circuits.*

Each complete cycle consists of two parts—the HIGH time and the LOW time. These two times are calculated separately. In the following discussion, we will assume the following values for the three frequency-determining components:

C1   50 μF
R1   12K
R2   3.3K

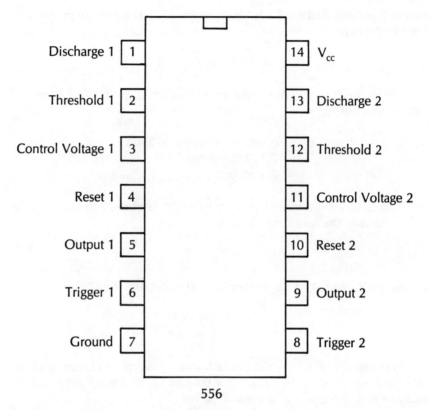

Fig. 10-40 *The 556 is a dual 555 timer.*

The time-per-cycle the output is in the HIGH state is determined by all three component values, C_1, R_1, and R_2, according to this formula:

$$T_h = 0.693C_1 1 \times (R_1 + R_2)$$

The value 0.693 is a constant determined by the internal construction of the 555 timer. Fortunately, we don't have to concern ourselves with how this value is derived.

For the component values for our example, listed above, the HIGH time works out to:

$$
\begin{aligned}
T_h &= 0.693 \times 0.00005 \times (12000 + 3300) \\
&= 0.0000346 \times 15300 \\
&= 0.52938 \text{ second} \\
&= 0.5 \text{ second}
\end{aligned}
$$

The LOW portion of the cycle is calculated in a similar manner, except this time, resistor R1 is totally ignored. The equation for the LOW time is simply:

$$T_l = 0.693 \times C_1 \times R_2$$

Using the component values for our example, we find that the LOW time works out to:

$$
\begin{aligned}
T_l &= 0.693 \times 0.00005 \times 3300 \\
&= 0.114345 \text{ second} \\
&= 0.1 \text{ second}
\end{aligned}
$$

The total cycle time, obviously enough, is simply the sum of the HIGH time and the LOW time. That is:

$$T_t = T_h + T_l$$

For example, we get an approximate total cycle time of:

$$
\begin{aligned}
T_t &= 0.5 + 0.1 \\
&= 0.6 \text{ second}
\end{aligned}
$$

It is usually more practical to talk about a waveform in terms of frequency instead of cycle time. The frequency is simply equal to the reciprocal of the total cycle time. That is:

$$
\begin{aligned}
F &= 1/T_t \\
&= 1/(T_h + T_l) \\
&= 1/[0.693C_1(R_1 + R_2)] + (0.693C_1R_2) \\
&= 1/[0.693C_1(R_1 + 2R_2]
\end{aligned}
$$

Using the component values from our example, the signal frequency is approximately equal to:

$$
\begin{aligned}
F &= 1/0.6 \\
&= 1.667 \text{ Hz}
\end{aligned}
$$

This is a pretty low frequency for many applications, but it is a good choice for our purposes in this project. If one of the oscillators has too high a frequency, above 8 to 10 Hz, or so, its controlled lamp will appear to be continuously lit, although perhaps at a lower than normal intensity. Actually, it will be blinking on and off at a rate too fast for the human eye to distinguish the individual flashes. If you decide to experiment with alternative component values, which I always strongly recommend, be sure to use relatively large values for the frequency determining components to keep their output frequencies low.

There is one other component to consider in the oscillator circuits. This is capacitor C2. This capacitor is included to help stabilize the 555 timer IC. It will not be necessary in all cases, but it is cheap insurance against possible problems.

The value of this capacitor is not at all critical, and has no direct effect on the operation of the circuit. A 0.01 μF capacitor is a commonly available value which will do the job nicely. Anything from about 0.005 μF to 0.05 μF will do fine.

In the schematic diagrams, the parts numbers in parentheses are for the additional duplications of each circuit in the complete project. This is done to make the parts list clearer.

The control circuit used in this project doesn't really need any detailed discussion. We have been using this triac-based light dimmer circuit in several of the projects in this chapter. For complete information on this circuit, refer to Project 19, where it was first presented.

I hope you enjoy this novel circuit. It doesn't have much in the way of practical applications, but it is fun and interesting. This type of project is often called a "Do-Nothing Box." Projects of this nature tend to be surprisingly popular with electronics hobbyists.

PROJECT 43: COLOR ORGAN

Color organs were very popular in the late sixties and early seventies. Then they fell out of public favor for a while. In recent years, they seem to be making something of a comeback. So, our final project in this chapter is a color organ circuit.

A color organ is a circuit which flashes different colored lights in response to an audio signal, usually music. The lighting varies with the beat of the music, often creating fascinating patterns.

A block diagram for this fun project is shown in Fig. 10-41. An audio signal is fed into the input of the circuit. This signal is then fed through four band-pass filters.

A filter is a frequency-sensitive circuit specifically designed to pass some frequencies well, while blocking, or severely attenuating other frequencies. A band-pass filter, as the name suggests, passes any frequency component within a specific, predetermined range, or band, to the output.

Any frequency component which falls outside this passed band is theoretically blocked by the filter, and doesn't get through to the output. In practical filter circuits, the blocking effect won't be 100 percent effective, especially for frequencies which are close to the limits of the passed band.

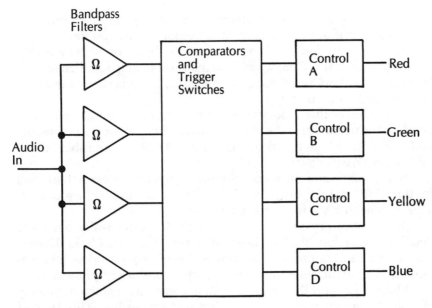

Fig. 10-41 *Color Organ—Block diagram*

The frequency response of a typical band-pass filter is illustrated in the graph of Fig. 10-42. There are two main specifications of concern for band-pass filters—the center frequency, and the bandwidth. Both of these terms are pretty much self-explanatory. The center frequency is the mid-point of the passed band, and the bandwidth indicates the overall size of the passed band.

In our color organ project, we are using the four band-pass filters to divide the audio spectrum into four more or less overlapping ranges. The bandwidths used here are fairly wide. We can identify the four ranges as follows:

- Bass
- Low Mid-Range
- High Mid-Range
- Treble

Each filter is followed by a voltage comparator. This is a circuit which compares two voltages and indicates at its output which voltage is the larger of the two.

One of the voltages for each comparator is a fixed value, called the reference voltage, or V_{ref}. The input voltage (the output from the preceding filter) is continuously compared to this reference voltage. Whenever the input voltage, or V_{in}, exceeds the value of the reference voltage, the

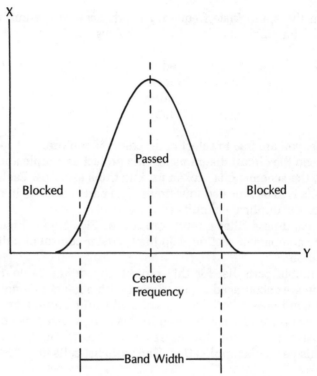

Fig. 10-42 *A band-pass filter passes some frequencies, while rejecting others.*

output of the comparator goes HIGH. If V_{in} is not larger than V_{ref}, the comparator's output will be LOW.

The comparator's output is used to control one of the four switches in IC3, a CD4066 quad bilateral switch. A HIGH control signal closes the appropriate switch.

Each switch controls an individual thyristor control circuit. When the switch is open, the thyristor cannot be triggered, so the output load remains OFF. When the control switch is closed, the thyristor circuit functions normally, triggering at an appropriate point in each input power cycle, as determined by the setting of potentiometer, R.

In this particular application, a miniature, screwdriver set, trimpot may be used, or the manual control could be replaced with an appropriate fixed resistor. This resistance value, whether fixed or adjustable, determines how much of the input power will reach the thyristor's load device, a colored lamp in this case, and therefore, how brightly the light will glow when activated.

Each thyristor and its associated load lamp can be activated only when the input filters and comparators determine that there is adequate energy in the appropriate frequency band. Each of the output lamps should be of a different color. Typical choices would be:

> red
> green
> yellow
> blue

Of course, you are free to select other colors if you prefer.

To keep the circuit diagrams for this project as simple and clear as possible, the schematic is broken up into three sections. Each of these subcircuits is to be repeated four times. The parts numbers in parentheses are for the duplicate circuit sections.

The band-pass filter circuit appears in Fig. 10-43. Figure 10-44 shows the comparator section, and the thyristor control circuit appears in Fig. 10-45.

A suitable parts list for this project appears as Table 10-29. As usual, you are encouraged to experiment with other component values.

The band-pass filter circuits are each built around an op amp. Almost any op amp IC may be used in this circuit. There is no need for an expensive, high-grade, low-noise device. Precise operation is irrelevant in this particular application. The parts list calls for a 324 quad op

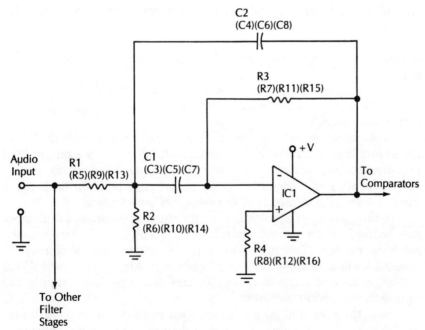

Fig. 10-43 *Color Organ—Band-pass filter circuit*

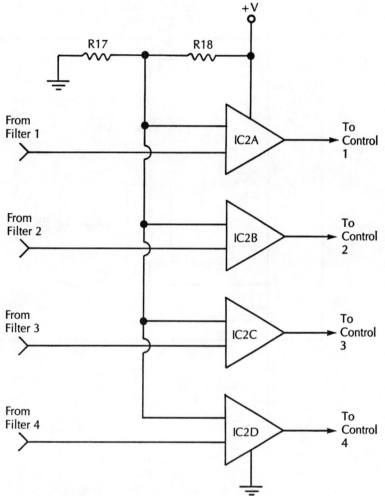

Fig. 10-44   *Color Organ—Comparator circuit*

amp. This chip is recommended because it is inexpensive and widely available. In addition, being a quad op amp, all four filters can be constructed around a single IC.

If you use another type of op amp IC, check the manufacturer's spec sheet carefully. Many op amp devices require a dual polarity power supply. Other than possible differences in the power supply requirements, virtually any op amp IC should work fine in this circuit without any other modifications.

To avoid confusion with the various duplicate circuit sections, the IC's pin numbers are omitted from the schematic diagram. The pin-out diagram for the 324 quad op amp IC appears in Fig. 10-45. If you use a

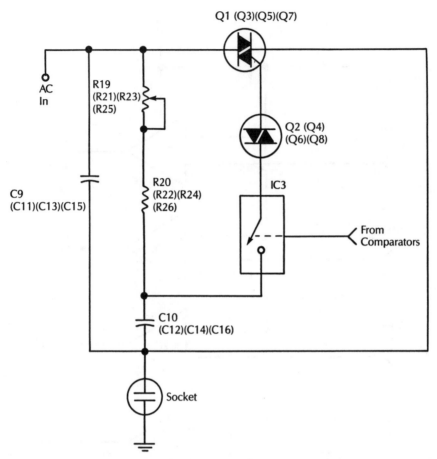

Fig. 10-45 *Color Organ—Thyristor control circuit*

different op amp chip, check the manufacturer's spec sheet for the correct pin-out numbers.

Both capacitors in this circuit should have identical values. That is:

$$C_1 = C_2$$

This is an active filter circuit, so there is some gain (amplification) for signals within the passed band of frequencies. The op amp's gain in this circuit is determined by resistors R1 and R3, according to this simple formula:

$$G = R_3/2R_1$$

Table 10-29   Parts List for Color Organ

IC1, IC2	324 quad op amp (xx)
IC3	CD4066 quad bilateral switch
Q1, Q3, Q5, Q7	Triac to suit load
Q2, Q4, Q6, Q8	Diac to suit triac
C1 - C8	0.1 μF capacitor
C9 - C16	0.1 μF 250 volt capacitor
R1	22K $^1/_4$ watt resistor
R2	47K $^1/_4$ watt resistor
R3	1.2K $^1/_4$ watt resistor
R4	33K $^1/_4$ watt resistor
R5, R12	8.2K $^1/_4$ watt resistor
R6	15K $^1/_4$ watt resistor
R7	470 ohm $^1/_4$ watt resistor
R8	12K $^1/_4$ watt resistor
R9, R16	4.7K $^1/_4$ watt resistor
R10	10K $^1/_4$ watt resistor
R11	270 ohm $^1/_4$ watt resistor
R13	3.3K $^1/_4$ watt resistor
R14	6.8K $^1/_4$ watt resistor
R15	180 ohm $^1/_4$ watt resistor
R17	100K $^1/_4$ watt resistor
R18	330K $^1/_4$ watt resistor
R19, R21, R23, R25	250K potentiometer
R20, R22, R24, R26	3.3K $^1/_2$ watt resistor

The filter's center frequency, F_c can be determined by using this equation:

$$F_c = 0.159 \times \sqrt{1/(R_3 \times 2C_1)} \times (1/R_1 + 1/R_2)$$

To set the bandwidth of the filter, we need to use a special quantity known as Q. Q is the ratio between the center frequency and the desired bandwidth. That is:

$$Q = F_c/BW$$

You can algebraically combine and rearrange these equations to find the three basic resistor values (R_1, R_2, and R_3) for the circuit:

$$R1 = Q/(2\pi \times F_c \times G \times C_1)$$

(π, or pi, is the mathematical constant, approximately equal to 3.14)

$$R_2 = 2G \times R_1$$

$$R_3 = Q/(2\pi \times F_c \times C_1 \times (2Q^2 - G)$$

The component values given in the parts list roughly divide the audio spectrum, from approximately 50 Hz to 15 kHz, into four slightly overlapping bands.

The next circuit section is the comparator stage, illustrated in Fig. 10-44. This part of the circuit is quite simple. A comparator is basically nothing more than an open loop op amp, wired as a difference amplifier with theoretically infinite gain.

Once again, almost any op amp IC may be used. The 324 quad op amp is a practical choice, because one chip can be used for all four comparator stages in the complete project.

The control circuit used in this project (Fig. 10-45) doesn't really need any detailed discussion. We have been using this triac-based light dimmer circuit in several of the projects in this chapter. For complete information on this circuit, refer to Project 19, where is was first presented.

This project may seem rather complicated at first glance, but it is really made up of a lot of redundant circuit elements, which are individually quite simple. You can have some fun with this project at your next party.

CHAPTER SUMMARY

In this chapter we have looked at over two dozen projects illustrating the versatility and wide range of potential applications for thyristors. I hope your imagination has been stirred to create your own projects.

❖11

The UJT and the PUT

SO FAR IN THIS BOOK WE HAVE BEEN STUDYING SCRs, TRIACS AND RELATED thyristors, which are intended primarily for power controlling applications. This is probably the most common type of thyristor.

But there is another class of thyristor devices, which we will explore in this chapter. The thyristors we are concerned with here are the *UJT* and the *PUT*.

WHAT IS A UJT?

The acronym *UJT* stands for *Unijunction Transistor*. To understand the significance of the name of this thyristor, we first need to consider the nature of an ordinary transistor.

A standard transistor is called a bipolar transistor because it has two active pn junctions. A UJT, as the name implies, effectively has only one active pn junction.

The standard schematic symbol used to represent a UJT is shown in Fig. 11-1. Compare this with the schematic symbol for a standard bipolar transistor, shown in Fig. 11-2. Both types of transistors have three leads. The three leads on a bipolar transistor are named:

- Emitter
- Base
- Collector

A UJT has no collector. Instead, it has two base leads. The terminals on a UJT are called:

- Emitter
- Base 1
- Base 2

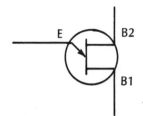

Fig. 11-1 *The UJT has just a single pn junction.*

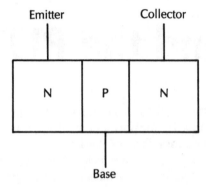

Fig. 11-2 *A standard bipolar transistor has two pn junctions.*

The reasons for this seemingly odd labeling scheme are made clear from the simplified diagram of a UJT's internal structure, illustrated in Fig. 11-3. This drawing is quite simplified for clarity. Unlike most thyristors, the UJT is a two-layer, rather than four-layer semiconductor device. Compared to other types of thyristors, UJTs are almost always low-power devices.

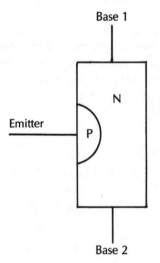

Fig. 11-3 *This is the internal structure of a UJT.*

Figure 11-4 shows a very rough equivalent circuit for a UJT. Electrically, the large n-type section acts like a resistive voltage divider, with a diode, the pn junction, connected to the common ends of the two resistance elements.

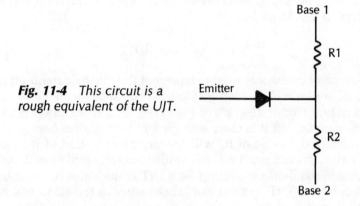

Fig. 11-4 *This circuit is a rough equivalent of the UJT.*

If a voltage is applied between base 1 and base 2, the diode will be reverse-biased. The emitter is assumed to be at ground potential, i.e., zero volts. Of course, this means that no current can flow from the emitter to either of the base terminals.

Now, let's suppose there is also an additional variable voltage source connected between the emitter and base 1, as illustrated in Fig. 11-5. As this emitter-base 1 voltage is increased from zero, a point will be reached when the diode becomes forward biased. Beyond this point, current can flow between the emitter and the bases. The UJT is "turned-ON" when the pn junction (the "diode") becomes forward-biased.

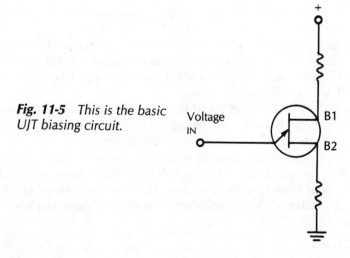

Fig. 11-5 *This is the basic UJT biasing circuit.*

THE INTRINSIC STAND-OFF RATIO

As with all semiconductor devices, there are a number of specifications used to define a UJT. Perhaps the most important single specification for a UJT is the *intrinsic stand-off ratio*. This is a resistance value. It is determined by the internal base resistances of the UJT, from Fig. 11-4, according to this formula:

$$\eta = R_a/(R_a + R_b)$$

η is the symbol commonly used to represent the intrinsic stand-off ratio. R_a and R_b are essentially identical to the resistance elements shown in the equivalent circuit diagram of Fig. 11-4. This description isn't 100 percent accurate, but it is close enough for our purposes here.

Typically the value of R_a will be very close to that of R_b. Consequently, the intrinsic stand-off ratio value is usually quite small. A typical intrinsic stand-off ratio rating for a UJT is the range of about 0.5 to 0.8. Most typical UJTs exhibit resistance values in the 4K to 10K range for both R_a and R_b.

Now let's return to the simple circuit of Fig. 11-5. Remember that as the emitter voltage is increased, eventually a point will be reached when the UJT's internal "diode" becomes forward-biased and permits current to flow from the emitter to the two bases. In order for this to happen, the emitter voltage must obviously exceed the voltage at the junction of the two bases, the common point between R_a, R_b, and the diode in Fig. 11-4.

The normal voltage at this junction is equal to the product of the voltage applied across the two bases times the intrinsic stand-off ratio. That is:

$$V_j = \eta(V_{bb})$$

If an ac, or varying voltage is applied to the emitter lead of a UJT, a series of pulses will appear across the bases. Figure 11-6 shows a typical ac UJT circuit. In this type of circuit, the output pulses are typically tapped off across resistor R2.

Resistor R1's chief function in this circuit is to improve the overall stability. This resistance keeps the circuit operating properly and consistently, despite any fluctuations in temperature.

An additional complimentary output may also be tapped off across resistor R1. The signal across resistor R1 is 180 degrees out of phase with the signal across resistor R2. That is, when the voltage across R2 goes more positive, the voltage across R1 will go more negative, and vice versa.

The values of resistors R1 and R2 are typically rather small. Usually they are significantly smaller than the UJT's internal resistances. This allows the circuit designer and experimenter to ignore the external resistances when analyzing the operation of the unijunction transistor.

UJT OPERATION

When an input pulse with enough amplitude to forward-bias the UJT's pn junction is applied to the emitter, output pulses, in step with the input pulses, will appear at the two bases, across resistors R1 and R2 in Fig. 11-6.

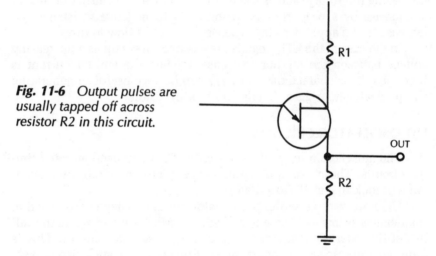

Fig. 11-6 *Output pulses are usually tapped off across resistor R2 in this circuit.*

In designing such a circuit, an important question to consider is, at what point does the pn junction actually become forward-biased and begin to conduct? This turn-ON value can be approximately defined with the following formula:

$$V_c = 0.5 + V_j$$

where V_c is the necessary conduction voltage and V_j is the UJT's internal junction voltage at the common point between internal resistances R_a and R_b.

If you remember the equation for the junction voltage (V_j) presented earlier in this chapter, you can see that this formula can be rewritten as:

$$V_c = 0.5 + \eta V_{bb}$$

where η is the intrinsic stand-off ratio and V_{bb} is the voltage applied across the UJT's two bases.

The UJT will be ON if the emitter voltage, V_e, is greater than the conduction voltage, V_c, defined above, and the UJT will conduct between the emitter and the bases. If the emitter voltage (V_e) is less than the conduction voltage, V_c, the UJT's pn junction will be reverse-biased and the emitter will not conduct. The UJT will be OFF.

Since a UJT has just a single pn junction, it can, under certain circumstances, exhibit the curious phenomenon known as "negative resistance." The concept of negative resistance was introduced in connection with four-layer diodes in chapter 1.

With negative resistance, Ohm's Law breaks down. Ordinarily, increasing the voltage across a specific resistance will cause the current to increase by a proportionate amount. With negative resistance, an increase in the applied voltage causes the current flow to drop.

In the case of the UJT, negative resistance shows up as a decreasing voltage between the emitter and base 1 when the emitter current is increased. This characteristic of UJTs can be very useful in generating sharp, clean gate-trigger signals for other thyristors.

UJT OSCILLATOR CIRCUITS

The most common applications for a UJT are in oscillator and timing circuits. These two applications are actually quite closely related. We will look first at UJT oscillators.

UJTs are very versatile in oscillator circuits. They can be used at frequencies as low as 1 Hz, and as high as 1 Mhz, depending on the values of the external components used in the oscillator circuit. This is quite an impressive operating range, especially for such a relatively simple circuit.

A very minor modification to the UJT demonstration circuit of Fig. 11-6 creates a simple but practical relaxation oscillator circuit. The modified circuit is illustrated in Fig. 11-7. The main modification here is the addition of a simple RC (resistor – capacitor) network in the emitter circuit.

When the power supply voltage, V_{bb}, is first applied to this circuit, the voltage across the capacitor, C, is obviously 0 volts. The capacitor hasn't been charged yet.

Once power has been applied to the circuit, the voltage across this capacitor will start to increase as it charges up from the supply voltage.

Notice that the voltage across capacitor C is directly applied to the emitter terminal of the UJT. The capacitor voltage is equal to the emitter voltage.

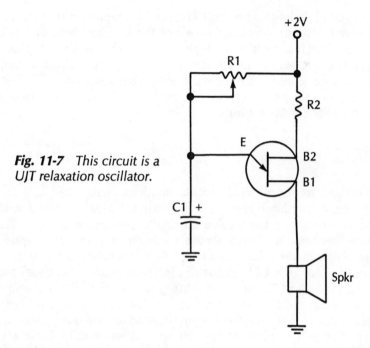

+2V

R1

R2

E

B2

B1

C1 +

Spkr

Fig. 11-7 *This circuit is a UJT relaxation oscillator.*

At some point, the voltage across capacitor C will exceed the V_c turn-ON point of the UJT. When this happens, the UJT will fire, generating an output pulse to its bases.

This firing of the UJT quickly discharges the capacitor, and the process then begins all over again. The cycle will be repeated endlessly, as long as power is supplied to the circuit, making this an oscillator circuit.

The time it takes capacitor C to reach the V_c point is dependent on several component values within the circuit. The most important component values here are those of capacitor C, resistor R3, and the intrinsic stand-off ratio of the specific UJT used in the circuit.

To simplify matters, the intrinsic stand-off ratio can usually be ignored in most circuits of this type, because its effect is negligible. This leaves us with capacitance C and resistance R3. The charging time can be considered approximately equal to:

$$T_c = CR_3$$

This formula is hardly exact, but it is close enough for most practical applications. Component tolerances are likely to introduce more practical error than the inherent inaccuracies of this simplified formula.

As stated earlier, when the UJT fires, the capacitor will quickly be discharged to ground through base 2 of the UJT and resistor R2. Then the next charging cycle will begin. The discharge time typically is extremely short. Again, its effect is generally quite negligible, and can reasonably be ignored in the circuit design calculations.

For practical purposes, we can calculate the oscillator frequency with this very simple formula:

$$F \;=\; 1/T_c$$
$$=\; 1/CR_3$$

This is about as simple as electronics math can get.

In most oscillator circuits it is most practical to select a likely capacitor value and then solve for the necessary resistance. This is because resistors are almost always available in more varied sizes and can easily be made adjustable using a potentiometer, or trimpot.

In the case of a UJT oscillator circuit, however, the usual recommended procedure is reversed. In this type of circuit, the best approach is to initially select a resistor value and then solve for the necessary capacitance. This is because this circuit places special requirements on the resistor value. Ideally, the timing resistor should have a value between:

$$[V_{bb}(1 - \eta) \;-\; 0.5]/2I_p$$

and:

$$[2(V_{bb} - V_v)]/I_v$$

where I_p is the maximum current flowing from the emitter to base 1, V_v is the "valley voltage" (the emitter – base 1 voltage just after the UJT has started to conduct) and I_v is the "valley current" (the emitter – base 1 current just after the UJT has started to conduct). V_{bb}, of course, is the supply voltage, which is applied across the UJT's base 1 and base 2 terminals.

Most of these values can be found in the manufacturer's specification sheet for the individual device used. These specifications are not to be considered exact. Typical values are listed on the data sheet. The actual value of your individual UJT may vary somewhat around the nominal value.

If the specification sheet for the UJT you want to use in the circuit is not available, you can find a suitable resistance value experimentally. Even if the resistance does not quite meet the requirements described here, the circuit should still work, provided the resistance value isn't too far off, though probably not at its peak performance capabilities.

Let's try a typical UJT oscillator circuit design example. Our goal here is to design a UJT relaxation oscillator with an output frequency of 750 Hz. The output signal amplitude, across resistor R2, should be 2.5 volts peak-to-peak.

To get started, we will assume the UJT we will be using in our circuit has the following specifications:

$$\eta \quad 0.52$$
$$I_p \quad 10 \ \mu\text{A}$$
$$V_v \quad 3.1 \text{ volts}$$
$$I_v \quad 25 \text{ mA}$$

Total internal resistance:

$$(R_a + R_b) \text{ 9000 ohms}$$

These values are fairly typical of most common UJTs.

The first step in designing our oscillator circuit is to select the circuit's supply voltage, V_{bb}. There is considerable leeway here, as long as the maximum voltage ratings of the UJT are not exceeded. In our sample circuit, we will use a supply voltage of 12 volts.

Next, we need to find the acceptable range of values for timing resistor R3. This resistor, as explained earlier, should have a value no lower than:

$$
\begin{aligned}
\text{R3}_{\text{min}} &= (2V_{bb} - V_v)/I_v \\
&= [2(12 - 3.1)]/0.025 \\
&= (2 \times 8.9)/0.025 \\
&= 17.8/0.025 \\
&= 712 \text{ ohms (minimum)}
\end{aligned}
$$

and no higher than:

$$
\begin{aligned}
\text{R3}_{\text{max}} &= [V_{bb}(1 - \eta) - 0.5]/2I_p \\
&= [12(1 - 0.52) - 0.5]/(2 \times 0.00001) \\
&= [12(0.48) - 0.5]/0.00002 \\
&= (5.76 - 0.5)/0.00002 \\
&= 5.21/0.00002 \\
&= 260{,}500 \text{ ohms (260.5K)}
\end{aligned}
$$

As you can see, these restrictions don't really limit us very much.

We have quite a range of acceptable values for resistor R3—anything from 712 ohms up to 260,500 ohms.

We will arbitrarily select a value of 22K for resistor R3. This would mean capacitor C needs a value equal to:

$$
\begin{aligned}
C &= 1/FR_3 \\
&= 1/(750 \times 22000) \\
&= 1/16500000 \\
&= 0.000000061 \text{ farad} \\
&= 0.061 \ \mu F
\end{aligned}
$$

Unfortunately, this is not a standard capacitance value. We could use a 0.062 μF capacitor, and it should be close enough. But for the sake of argument, let's say we can't get a 0.062 μF capacitor. In this case, we will have to settle for the nearest available value. Let's say the best we can do is use a 0.047 μF capacitor. This changes the output frequency to:

$$
\begin{aligned}
F &= 1/CR_3 \\
&= 1/(0.000000047 \times 22000) \\
&= 1/0.001034 \\
&= 967 \text{ Hz}
\end{aligned}
$$

This isn't very close to our target output frequency of 750 Hz.

Since we have found a capacitance value that almost fits, we can now go back and calculate a new value for resistor R3 that should still be well within the circuit's acceptable range:

$$
\begin{aligned}
R_3 &= 1/FC \\
&= 1/(750 \times 0.000000047) \\
&= 1/0.000353 \\
&= 28,369 \text{ ohms}
\end{aligned}
$$

The nearest standard resistance value is 27K. This is still well within the acceptable range of resistance values of 712 ohms to 260,500 ohms. A 27K resistor for R3 will probably work just fine.

But, since we've rounded off the resistor value, the output frequency will be altered. Assuming perfect 0 percent tolerance components, the actual output frequency will be approximately:

$$
\begin{aligned}
F \ & 1/CR_3 \\
&= 1/(0.000000047 \times 27000) \\
&= 1/0.001269 \\
&= 788 \text{ Hz}
\end{aligned}
$$

We have an error of 38 Hz. This should be close enough for most practical applications. After all, the error tolerances of practical components could cause at least this much error.

If your individual application calls for a more precise output frequency, reduce the value of resistor R3 somewhat and add a trimpot in series. In our sample circuit, we could use a 22K resistor in series with a 10K trimpot.

This will allow us to change the total R3 resistance from a low of 22000 ohms to a high of 32000 ohms. By monitoring the output signal with an accurate oscilloscope or frequency counter, we can tune the oscillator to the exact desired output frequency by adjusting the shaft of the trimpot.

By permitting such a wide tuning range, we have plenty of "elbow room" on either side of the required resistance value to more than compensate for any possible component tolerance errors.

The bulk of our design is already done, but we are not quite finished with our oscillator circuit yet. What values should resistors R1 and R2 have in our circuit?

The equation for finding the desired value of resistor R1 is:

$$R_1 = (R_a + R_b)/(2\ \eta\ V_{bb})$$

For our sample circuit design, this works out to a resistance of:

$$
\begin{aligned}
R_1 &= 9000/(2 \times 0.52 \times 12) \\
&= 9000/12.48 \\
&= 721\ \text{ohms}
\end{aligned}
$$

In most UJT oscillator circuits, resistor R1 will usually have a value somewhere between 500 ohms and 1000 ohms. This resistance is not particularly critical. We can round it off to the nearest standard value available. In this case, we can use either a 680-ohm or a 750-ohm resistor, and the circuit will work fine. In our sample circuit, we will use a 680-ohm resistor for R1.

In fact, if your individual application isn't particularly critical, you can go ahead and use something like a 680-ohm or 750-ohm resistor for R1 in all oscillator circuits of this type. You don't really have to bother with the equation at all. As above, R1 will normally have a value between 500 ohms and 1000 ohms, so you can usually estimate, and use a resistance somewhere in the middle of this range.

If you don't calculate the correct value for resistor R1, the amplitude of the output signal might not be quite what you intended, but it should be close.

Now, all we have to do is find the correct value for resistor R2. The formula is:

$$R_2 = \{[(R_a + R_b) + R1]V_o\}/(V_{bb} - V_o)$$

where V_o is the desired peak-to-peak output voltage across resistor R2.

In our sample design circuit, the ideal value for resistor R2 is:

$$
\begin{aligned}
R_2 &= [(9000 + 680)2.5]/(12 - 2.5) \\
&= (9680 \times 2.5)/9.5 \\
&= 24200/9.5 \\
&= 2547 \text{ ohms}
\end{aligned}
$$

Again, unless the intended application is extremely critical, we can round off the value of resistor R2. We can use either a 2.2K or a 2.7K resistor here. This completes our design of a typical UJT oscillator circuit.

UJT TIMING CIRCUITS

A UJT timing circuit is very similar to a UJT oscillator circuit. The major difference here is that there is no feedback path to automatically recharge the capacitor after the UJT fires. A momentary action switch initializes the timing cycle by charging the capacitor.

After the timing cycle is completed, the capacitor discharges and the UJT fires. There is one and only one output pulse each time the circuit is activated, rather than a continuing series of output pulses, as in an oscillator circuit.

A timer circuit is usually set up for a longer time period (capacitor charging time) than a typical oscillator. This just means that larger values for capacitor C and timing resistor R3 are employed in the timer circuit.

THE PUT

Functionally related to the UJT is the *Programmable Unijunction Transistor* or *PUT*. Like most thyristors, the PUT is a four-layer semiconductor device. The basic internal structure of a PUT is illustrated in Fig. 11-8. The PUT has three leads, like most thyristors. They are labeled:

- A   anode
- C   cathode
- G   gate

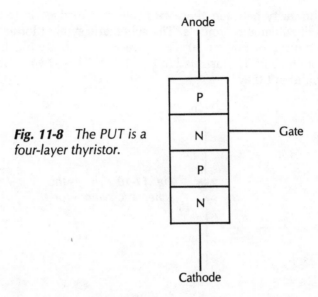

Fig. 11-8 *The PUT is a four-layer thyristor.*

From the information presented so far, you might suspect that the PUT has a lot in common with the SCR. (Refer to chapter 2.) If this is what you are thinking, you're right. We can readily see the similarity if we compare the internal structure of a PUT with that of an SCR. The internal structure of an SCR is shown again in Fig. 11-9. The only important structural difference between the PUT and the SCR is the placement of the gate terminal.

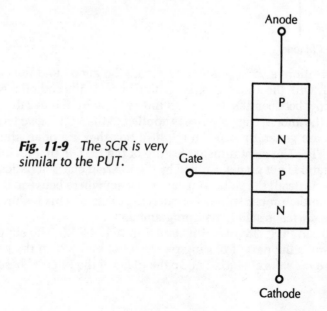

Fig. 11-9 *The SCR is very similar to the PUT.*

The similarity between these two thyristor devices is re-emphasized in their schematic symbols. The schematic symbol for an SCR is shown once more in Fig. 11-10. Compare this symbol with the schematic symbol for a PUT, illustrated in Fig. 11-11. These two symbols are very similar, aren't they?

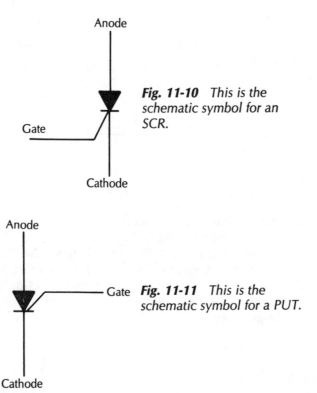

Fig. 11-10   *This is the schematic symbol for an SCR.*

Fig. 11-11   *This is the schematic symbol for a PUT.*

In operation, a voltage is placed across the anode and the cathode of the PUT, with the anode positive with respect to the cathode. No current can flow between the anode and the cathode until a negative (with respect to the anode) trigger pulse is applied to the PUT's gate terminal.

PUTs are generally used in applications that are often similar to those of UJTs. The most unusual feature of the PUT is that its intrinsic stand-off ratio is not predetermined by the internal characteristics of the PUT itself. Instead, the value of n can be set anywhere between 0 and 1, by selecting the proper values for external resistors. This feature is the reason the device is said to be "programmable."

The basic PUT circuit is illustrated in Fig. 11-12. The significant feature here is the pair of biasing resistors, R1 and R2, in the gate circuit. These resistors essentially take the place of the internal base resis-

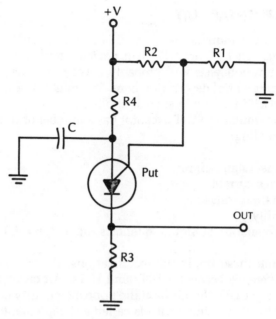

Fig. 11-12 *This is the basic PUT circuit.*

tances, R_{bb}, in a UJT. The resistor values replace the internal resistances in all circuit calculations. The intrinsic stand-off ratio, η, is determined by this ratio:

$$\eta = R_1/(R_1 + R_2)$$

Other than the programmable intrinsic stand-off ratio, all of the calculations for a PUT circuit are the same as for a similar UJT circuit.

The I_p (peak current) rating for a PUT is usually significantly lower than the I_p rating for a comparable UJT. This results in a much lower leakage current. In a well-designed PUT circuit, the leakage current can be as low as 0.1 mA. This low voltage leakage current can permit long delays between output pulses. If a PUT is used in an oscillator circuit, it can operate with a much lower frequency. In a timer circuit, the PUT can offer longer time delays than a UJT circuit.

When a PUT is forward-biased, the voltage drop between the anode and the cathode is very low. This makes the PUT a very efficient switching device.

The I_v, "valley" current, rating for a PUT also tends to be lower than for a UJT. This allows PUT oscillator circuits to operate from a lower supply voltage.

THE COMPLEMENTARY UJT

Another semiconductor device that is closely related to the UJT is the *Complementary UJT*, or *CUJT*. This device is fairly uncommon, and most electronics hobbyists will probably never run across one. Still, I think it warrants a brief description here. You might find it worthwhile to seek out a CUJT for certain special applications.

A CUJT is similar to a UJT except it has a number of superior specifications, including:

- lower intrinsic stand-off ratio
- lower leakage current
- lower saturation voltage
- better stability
- capable of operation at high frequencies (up to 100 kHz).

Aside from these improved specifications, there is another very important difference between CUJTs and UJTs. All currents and voltages applied to a CUJT should have the opposite polarity of those used with a UJT. This is why the device is called a *Complementary UJT*. All of the signals are reversed. CUJTs are used in applications similar to UJT circuits.

PROJECT 44: SAWTOOTH-WAVE OSCILLATOR

The remainder of this chapter will feature a few more projects, all using UJTs. The circuit for our first UJT project appears in Fig. 11-13. A typical parts list for this project is given in Table 11-1.

This project is an oscillator circuit, but one that is a little more sophisticated than the basic UJT oscillator circuit discussed earlier in this chapter. The main output signal of this circuit is a sawtooth, or ramp waveform. The output signal from this project is illustrated in Fig. 11-14.

Actually, Fig. 11-14 is a rather idealized version of this circuit's output. This circuit really only generates a rather rough approximation of a sawtooth wave. Still, it should be close enough for most noncritical applications. The values of capacitors C1 and C2 should be as closely matched as possible. An accurate capacitance meter would be a big help in calibrating this project.

The more precisely matched these two capacitor values are, the more linear (true sawtooth-like) the output signal will be. In semicritical applications, this circuit can be used with precision capacitors.

Transistors Q1 and Q2 are UJTs. Transistor Q3 is a Field Effect Transistor (FET). Other type numbers may be substituted in place of the units called for in the parts list, but the operating parameters of the cir-

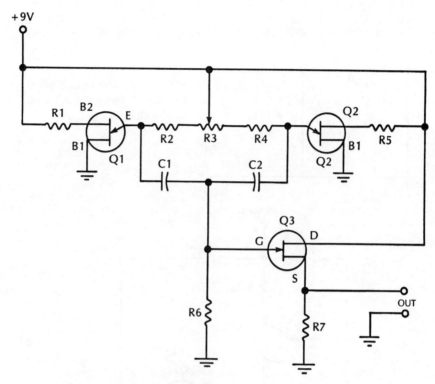

Fig. 11-13   *Sawtooth-Wave Oscillator*

Table 11-1   Parts List for Sawtooth-Wave Oscillator

Q1, Q2	UJT (TIS-43, or similar)
Q3	FET (MPF103, or similar)
C1, C2	0.01 μF capacitor
R1, R5, R7	4.7K $^1/_4$ watt resistor
R2, R4	100K $^1/_4$ watt resistor
R3	1 Megohm potentiometer
R6	1 Megohm $^1/_4$ watt resistor

cuit may be altered unless the substitute device is very similar to the original. Overall, this circuit isn't too picky about the exact devices used. Potentiometer R3 is used to manually adjust the output frequency.

PROJECT 45: SIMPLE VCO

The circuit shown in Fig. 11-15 is a *Voltage Controlled Oscillator*, or *VCO*. The output frequency can be changed by applying an input

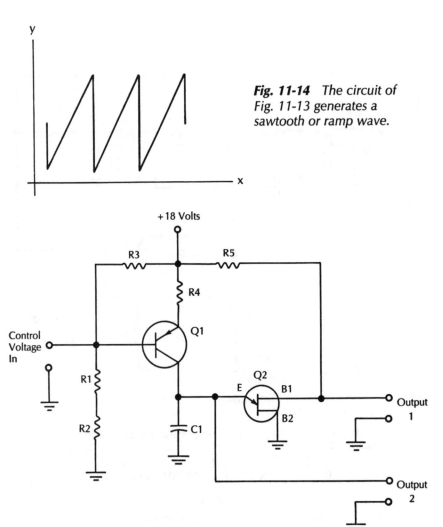

Fig. 11-14 *The circuit of Fig. 11-13 generates a sawtooth or ramp wave.*

Fig. 11-15 *Simple VCO*

voltage. The input voltage is called, naturally enough, the "control voltage."

VCOs are used in many applications, including electronic music, electronic tuning, automation systems, phase-locked loops, (PLLs), and others.

To operate properly, this circuit must be supplied by a +18 volt power source. Other than that, nothing is terribly critical here. Almost any low-power pnp transistor may be used for Q1.

A complete suggested parts list for this project appears as Table 11-2. Feel free to experiment with different component values.

Table 11-2   Parts List for Simple VCO

Q1	PNP transistor (2N2393, Radio Shack RS2021, or similar)
Q2	UJT (2N441, 2N2077, Radio Shack RS2006, or similar)
C1	0.022 μF capacitor
R1	6.8K $^1/_4$ watt resistor
R2	680 ohm $^1/_4$ watt resistor
R3, R4	2.2K $^1/_4$ watt resistor
R5	330 ohm $^1/_4$ watt resistor

The circuit has two independent outputs. They may be used separately, but remember, the signals at both outputs will always, by definition, have identical frequencies. The signal at output 1 is a fair approximation of a sawtooth wave (refer to Fig. 11-14). Output 2 puts out a negative spike wave, like the one illustrated in Fig. 11-16.

PROJECT 46: MULTI-WAVEFORM GENERATOR

The circuit shown in Fig. 11-17 puts out three different waveforms. Such a multi-waveform generator is sometimes called a "function generator." It is a very useful device for electronics hobbyists and technicians. In one compact project, we have several common waveforms to work with. This circuit can provide the needed signal for most audio applications, including testing. A suitable parts list for this project is given in Table 11-3. This project can generate three different waveforms. The waveform at the output is selected by switch S2. The three available waveforms are:

- triangular wave
- square wave
- sawtooth wave.

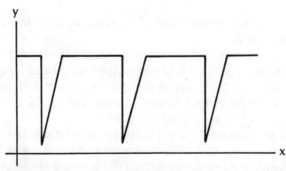

Fig. 11-16 *The secondary output of the circuit shown in Fig. 11-15 is a negative spike wave.*

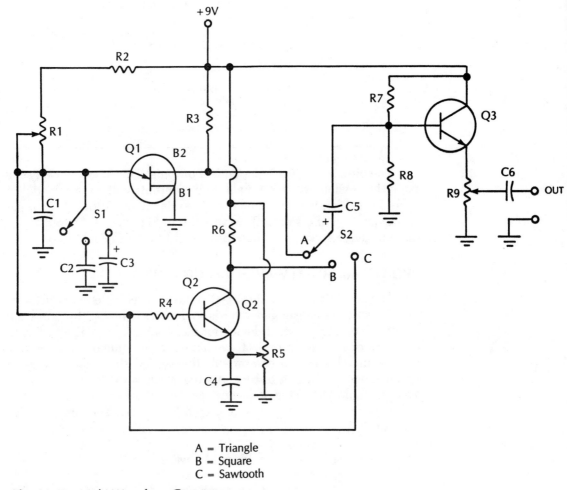

A = Triangle
B = Square
C = Sawtooth

Fig. 11-17 *Multi-Waveform Generator*

These three signal types are probably the most commonly used waveforms in electronics.

One disadvantage of this circuit is that the three waveforms are not
simultaneously available. Only one output signal can be generated by
the circuit at a time. This may or may not be a limitation, depending on
your specific intended application. In most testing, for example, you'll
only be interested in one signal at a time anyway.

Switch S1 is a range control for the project. It sets the overall range
of output frequencies that can be generated. Potentiometer R2 is used to
fine tune the output frequency within the selected range. As the project
is shown here, the circuit has three possible ranges. You could add

Table 11-3   Parts List for Multi-Waveform Generator

Q1	UJT (2N2646, Radio Shack RS2029, or similar)
Q2, Q3	PNP transistor (2N4124, GE-20, or similar)
C1	0.01 μF capacitor
C2	0.1 μF capacitor
C3	1 μF 25 volt electrolytic capacitor
C4, C5	47 μF 25 volt electrolytic capacitor
C6	0.47 μF capacitor
R1	25K potentiometer (frequency)
R2	2.2K $1/4$ watt resistor
R3	1.5K $1/4$ watt resistor
R4, R6	10K $1/4$ watt resistor
R5	50K potentiometer (distortion)
R7, R8	100K $1/4$ watt resistor
R9	5K potentiometer (signal amplitude)
S1, S2	Single-Pole 3 Position rotary switch

more ranges or reduce the number of ranges easily enough. The three ranges in this version of the circuit are set up to overlap one another.

Depending on the accuracy of your component values, the overall range of output frequencies for this circuit should run from a low of about 200 Hz up to a high of approximately 20 kHz. The operating range of this circuit cannot be easily increased, even if you add more range selections to switch S1. All switch S1 can do is break up the circuit's overall frequency range into more convenient "hunks."

In addition to the two switches, this circuit has three more controls. Potentiometer R1 sets the actual output frequency within the selected range. The waveshape can be fine-tuned and distortion minimized by adjusting potentiometer R5. It may be a good idea to use a trimpot for this control. Finally, potentiometer R9 is more or less a volume control. It sets the output amplitude or signal level.

PROJECT 47: TIMER

Figure 11-18 shows how a UJT can be used in a practical thyristor timer circuit. The circuit turns on an SCR and its load for a predetermined amount of time.

Note that this circuit is intended for dc operation only. Do not attempt to drive an ac load with this device. A typical parts list for this project is given in Table 11-4.

SCR Q1 turns the load ON and SCR Q2 turns it OFF after the circuit has timed out. Almost any SCRs that can handle the power required by the load device will work in this circuit.

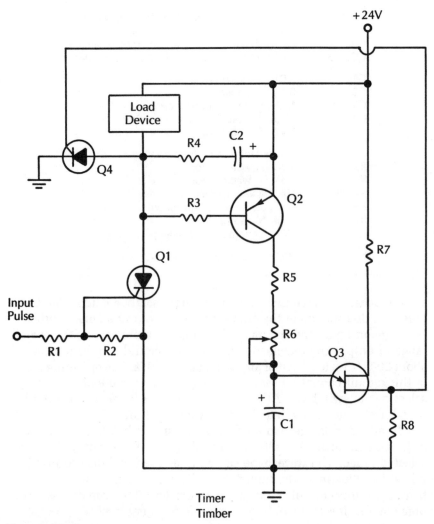

Fig. 11-18 Timer

The time constant of the timer circuit is determined by the values of capacitor C1 and resistors R5 and R6. R6 is a potentiometer that permits the operator to set different timing values. It is a good idea to calibrate this control. Use a stopwatch to determine the settings for various times, and mark their positions on the knob.

You can also alter the timing period of this circuit by changing the value of capacitor C1. The larger this capacitor is, the longer the time period will be.

Table 11-4   Parts List for Timer

Q1, Q4	SCR (2N5061, or similar—see text)
Q2	PNP transistor (2N4125, or similar)
Q3	UJT (2N4853, or similar)
C1	2.2 μF 50 volt electrolytic capacitor
C2	1 μF 50 volt electrolytic capacitor
R1	100 ohm 1/2 watt resistor
R2, R7	1K 1/2 watt resistor
R3, R4	10K 1/2 watt resistor
R5	22K 1/2 watt resistor
R6	5 Megohm potentiometer
R8	27 ohm 1/2 watt resistor

PROJECT 48: CASCADED TIMER

Two or more of the timer circuits of Fig. 11-18 can be combined to create a cascaded, or sequential timer. A two-stage cascaded timer circuit is illustrated in Fig. 11-19. Compare this circuit carefully to the one shown in Fig. 11-18. Notice how similar these two circuits are.

In operation, when the circuit is triggered by an input pulse, load A will be turned ON, via SCR Q1. After a time constant, determined by capacitor C1, and resistors R4 and R5, load B will be turned ON, via SCR Q4. After a second time constant, set by capacitor C4 and resistors R9 and R10, both of the load devices will be turned back OFF by SCR Q7. UJTs Q3 and Q6 are the primary timing devices in this circuit. Additional timer stages can easily be added to this circuit. If only load B is used, there will be a delay equal to the first time constant after the triggering input pulse before the load device is turned ON. This may be desirable in certain automation-type applications. A suitable parts list for this project appears as Table 11-5. Experiment with other values for the timing components:

- Time Period A

    capacitor C1
    resistor R4
    potentiometer R5

- Time Period B

    capacitor C4
    resistor R9
    potentiometer R10

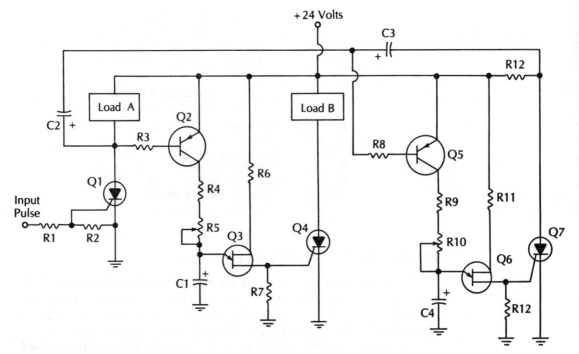

Fig. 11-19 Cascaded Timer

Table 11-5   Parts List for Cascaded Timer

Q1, Q4, Q7	SCR (2N5061, or similar—see text)
Q2, Q5	PNP transistor (2N4125, or similar)
Q3, Q6	UJT (2N4853, or similar)
C1 - C4	1 μF 50 volt electrolytic capacitor
R1	100 ohm $^1/_2$ watt resistor
R2, R6, R11	2.2K $^1/_2$ watt resistor
R3, R8, R12	10K $^1/_2$ watt resistor
R4, R9	22K $^1/_2$ watt resistor
R5, R10	5 Megohm potentiometer
R7, R12	27 ohm $^1/_2$ watt resistor

Of course, the potentiometers give manual control over the timing periods. Changing the capacitor values will change the range of time periods for each stage. The two stages do not have to be set for equal timing periods. Different timing capacitor values may be used in each individual stage.

This project is designed to drive dc loads only. Do not attempt to control any ac powered device with this circuit.

PROJECT 49: WIDE RANGE TIMER

A somewhat different UJT timing circuit is illustrated in Fig. 11-20. This circuit is intended to control low-power loads. A typical parts list for this project is given in Table 11-6.

Momentary action, normally open, SPST switch S1 is used to trigger the timer circuit. Switch S2 is an optional range selector switch.

The timing period of this circuit is determined by the values of capacitor C1, resistor R2, potentiometer R3, and any resistors selected by the range switch, S2.

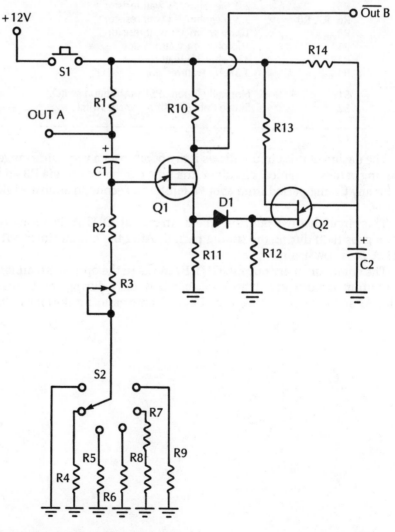

Fig. 11-20   *Wide-Range Timer*

Table 11-6 Parts List for Wide Range Timer

Q1, Q2	UJT (2N2646, or similar)
D1	almost any silicon diode
C1	2.2 μF 25 volt electrolytic capacitor
C2	0.22 μF capacitor
R1, R10, R12	47 ohm $1/4$ watt resistor
R2	330K $1/4$ watt resistor
R3	5 Megohm potentiometer
R4	1 Megohm $1/4$ watt resistor
R5	2.2 Megohm $1/4$ watt resistor
R6, R7, R8	3.3 Megohm $1/4$ watt resistor
R9	10 Megohm $1/4$ watt resistor
R11	100 ohm $1/4$ watt resistor
R13	330 ohm $1/4$ watt resistor
R14	3.3K $1/4$ watt resistor
S1	Normally Open SPST pushbutton switch
S2	Single Pole 6-Position rotary switch

The design of this circuit allows it to operate over a very wide range of timing values. A typical circuit of this type can be set up, via R3 and S2, for any timing period from about a tenth of a second to almost a full minute.

This circuit has two outputs. The main output, OUT A, is normally LOW. It goes HIGH during the timing period. After the circuit times out, OUT A goes LOW again.

The complementary output, OUT B, works in the opposite manner. This output is normally HIGH, and goes LOW only during the timing cycle. After the circuit times out, OUT B returns to its normal HIGH state.

❖ 12
Specialized Devices

IN THIS FINAL CHAPTER, WE WILL TAKE A QUICK LOOK AT A FEW LESS COMMON thyristor devices. These thyristors are generally rather rare, especially on the hobbyist market, but you may possibly encounter some of them from time to time in your electronics work. The thyristors covered in this chapter are nowhere near as common as SCRs, triacs, diacs, or UJTs, but they are still important.

Thyristor types covered in this chapter include:

- CSCR   Complimentary Silicon Controlled Rectifier
- GSC   Gate Controlled Switch
- SBS   Silicon Bilateral Switch
- SCS   Silicon Controlled Switch
- SUS   Silicon Unilateral Switch
- LASCR   Light Activated SCR

Some of these device names are self-explanatory.

THE CSCR

The first of the thyristor devices we will consider in this chapter is the *Complementary Silicon Rectifier* or *CSCR*. In the simplest terms, a CSCR is a negatively triggered SCR.

Where an ordinary SCR requires a positive pulse on its gate terminal to turn ON, a CSCR requires a negative gate pulse. Otherwise, the two devices are quite similar and are used in many of the same applications. The CSCR is useful when you are using a trigger signal from some pre-existing circuitry and the trigger signal happens to be negative rather than positive.

By an unfortunate coincidence, the same schematic symbol is used for both the CSCR and the PUT. This could cause some confusion. Check the parts list carefully to determine which device is to be used.

The schematic symbol used for both the CSCR and the PUT is illustrated in Fig. 12-1. Aside from the fact that both are thyristors, the CSCR and the PUT have very little in common, either in terms of internal structure or in function.

Fortunately, CSCRs and PUTs are seldom used in the same circuit, so the overly similar schematic symbols shouldn't be too much of a problem in most cases.

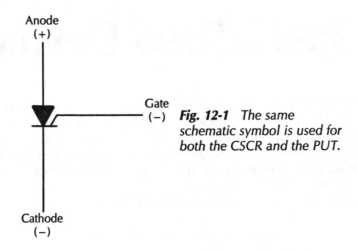

Fig. 12-1 *The same schematic symbol is used for both the CSCR and the PUT.*

THE GSC

The *Gate Controlled Switch*, or *GSC* is sometimes called the *Gate Turn-Off Switch*, or *GTO*. The GSC was developed to overcome the chief limitation of SCRs. The gate terminal can be used to turn the SCR ON, but it cannot turn it OFF. The GSC, unlike the SCR, can be turned on or off via its gate lead.

If a positive pulse is applied to the gate terminal of a GSC, it functions like an ordinary SCR. A positive gate pulse turns the device ON. On the other hand, if the GSC is already ON and a negative pulse is applied to the gate terminal, the device will be turned OFF. There is no need to remove the anode current, as is the case with ordinary SCRs.

To summarize the operation of a GSC, a positive gate pulse turns the device ON, and a negative gate pulse turns it OFF. The internal structure of a GSC is illustrated in Fig. 12-2. The schematic symbol for this device is shown in Fig. 12-3.

The operating times of the GSC are quite fast. The device can be electrically turned ON or turned OFF in a matter of a few microseconds.

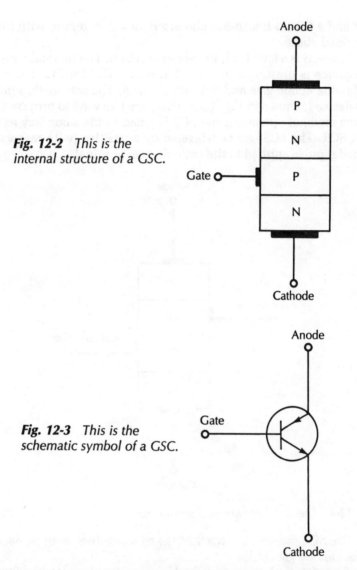

Fig. 12-2 *This is the internal structure of a GSC.*

Fig. 12-3 *This is the schematic symbol of a GSC.*

As with everything else in life and electronics, there are certain trade-offs involved in using a GSC. Compared to a standard SCR, the GSC has rather inferior maximum current and dissipation ratings. In addition, a GSC requires a higher gate current to turn the device ON and OFF.

THE SCS

The next thyristor device we will be looking at is the *Silicon Controlled Switch*, or SCS. The SCS is rather like a combination of both an

SCR and a CSCR. It also has characteristics in common with the GSC, discussed above.

The SCS is a four-lead, four-layer thyristor. The internal structure of this device is illustrated in Fig. 12-4. Notice that the SCS has two gate leads—the anode-gate and the cathode-gate. The schematic symbol for this device is shown in Fig. 12-5. There are two ways to turn ON an SCS. In one mode of operation, the SCS is gated in the same way as a standard SCR. The SCS can be triggered by a positive (with respect to the cathode) pulse applied to the cathode-gate terminal.

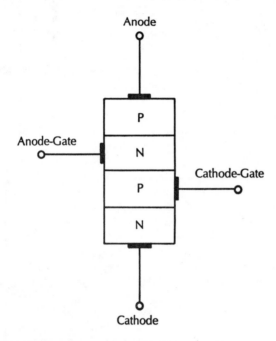

Fig. 12-4 *The SCS has two gate terminals.*

The SCS can also be switched ON by a negative (with respect to the anode) anode-gate pulse.

Turn-OFF can be accomplished by reversing the polarity of the turn-ON gate pulse for either mode of operation. A rough equivalent circuit for an SCS can be created from an npn/pnp pair of transistors, as shown in Fig. 12-6. We will use this diagram in the following discussion of the functioning of an SCS.

Initially, we will assume that the device is in its OFF state. No signal is applied to either of the gates. At this point, no current can flow between the anode and the cathode. The collector-base junction in each of the two component transistors is reverse-biased. The device is in an OFF state.

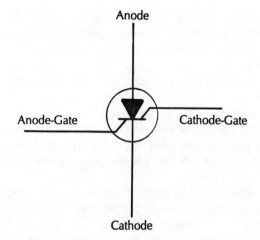

Fig. 12-5   *This is the schematic symbol for the SCS.*

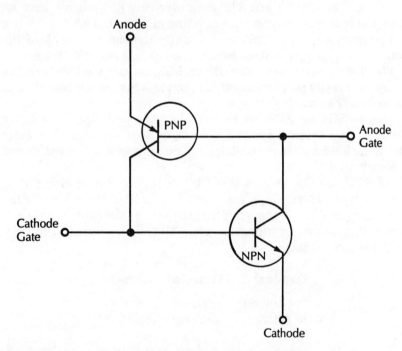

Fig. 12-6   *This is a rough equivalent circuit for an SCS.*

Any circuit using an SCS must avoid any dc circuit path from the anode-gate lead to ground. If the anode-gate is shorted to ground, load current can flow through the emitter-base junction of the upper (pnp) transistor.

The SCS, or the two transistor construct of Fig. 12-6, can be turned ON by applying a positive pulse to the cathode-gate base of the lower (npn) transistor. A base current in this transistor results in collector current. This forces current to flow through the base of the other upper pnp transistor. Regenerative action immediately latches both of the transistors into saturation.

Similarly, the SCS can also be switched from its OFF state to its ON state by applying a negative pulse to the anode-gate terminal. Because the two internal transistors have opposite polarities, we get the same results as above. Turning ON the SCS via either gate works in exactly the same way.

Generally speaking, the cathode-gate is more sensitive than the anode-gate. This means that a smaller positive current pulse at the cathode-gate is required to trigger the device. To trigger the SCS at the anode-gate terminal requires a larger negative current pulse.

For a typical SCS, the minimum trigger current at the cathode-gate will be rated at about 1 μA. The same device will probably have an anode-gate minimum trigger current rating of 80 to 120 μA.

The required trigger voltage is usually the same for both of the gates. A typical trigger voltage rating for an SCS is about 0.75 volt.

When it comes time to turn OFF an SCS, we can see how versatile this thyristor really is. There are three ways this device can be switched back to its OFF state.

Like an SCR, an SCS can be turned OFF by permitting the anode current to drop below a specific holding current (I_h) level. The exact value of the holding current will depend on the specific construction of the device used.

An SCS may also be turned OFF by a trigger pulse on either of its gates. The turn-OFF trigger pulses must have the opposite polarity of the turn-ON trigger pulses. A negative trigger pulse on the cathode-gate or a positive trigger pulse on the anode-gate will turn OFF the SCS. To summarize the gate signals:

Terminal	Turn-On	Turn-Off
Gate-Cathode	positive	negative
Gate-Anode	negative	positive

Take another look at the equivalent circuit shown in Fig. 12-6. Assume that both transistors are ON, as described above. If a negative trigger pulse is applied to the cathode-gate, the lower (npn) transistor will but cut OFF and it will stop conducting.

Because this transistor is no longer conducting, its collector current vanishes. Therefore, there is now no base current being supplied to the upper (pnp) transistor. This device is now also cut OFF. The SCS, as a unit, is back in its OFF state.

Exactly the same process occurs if a positive trigger pulse is fed to the anode-gate while the SUS is ON, except all signal polarities are simply reversed.

SCSs are normally used in low-power applications. The advantage of gate controlled turn-OFF is important for many specialized control circuits. Typical SCS applications include:

- Digital counters
- Lamp drivers
- Neon display drivers
- Oscillators
- Pulse generators
- Relay drivers
- Shift registers
- Voltage level sensors

among other applications.

THE SUS

The *Silicon Unilateral Switch*, or *SUS* is generally employed as a gate-trigger device for power-control elements. The schematic symbol for this device is shown in Fig. 12-7. The SUS is a relatively recent addition to the thyristor family. Its internal structure is rather complex.

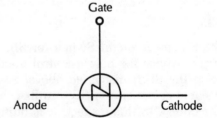

Fig. 12-7 *The SUS is usually employed as a gate-trigger device for power-control elements.*

In some ways, the SUS is rather like a polarized diac (see chapter 3). Unlike a diac, an SUS passes current in only one direction, as the term "unilateral" in the device name implies. Another difference between a diac and an SUS is that the SUS has a third "gate" lead, similar to the one on an SCR.

The SUS is a fairly new type of thyristor. It is far more sophisticated than similar devices such as diacs. Don't, however, assume too much from that last statement. The comparison I made between these two thyristor devices (the SUS and the diac) was purely in terms of function. Actually, an SUS has more in common with an integrated circuit, or IC.

Like most thyristors, the SUS is basically, a four-layer semiconductor device. The internal structure of an SUS is quite similar to that of an SCR. The primary difference is that the gate terminal of an SUS has a built-in zener diode, which is not included within a standard SCR.

By including this internal zener diode in the gate circuit, the overall sensitivity of the switching device is reduced. Small trigger pulses are completely ignored by the SUS. Most stray noise pulses won't cause any problems in circuit operation. Problems of false triggering due to noise pulses can show up in SCR circuits.

To trigger an SUS, the applied gate voltage must exceed a specific, predetermined level. The triggering voltage level for an SUS is significantly higher than the triggering voltage level for a comparable SCR. The breakdown voltage for an SUS is relatively low. A typical value is about 8 volts.

The current through an SUS may flow from anode to cathode only. This type of thyristor should not be used in any circuit which can reverse-bias the SUS. A cathode-to-anode current could easily damage or destroy the device.

An SUS is turned OFF in the same way as an SCR. The primary method of device turn-OFF is to reduce the anode current below the unit's holding current (I_h) level, as defined in the manufacturer's specification sheet.

Like an SCR, an SUS can go from an OFF state to an ON state faster than it can go from an ON state to an OFF state. The SUS is mainly used in high-speed switching circuits and similar applications.

THE SBS

The *Silicon Bilateral Switch*, or *SBS* is generally employed as a gate-trigger device for power-control elements. This thyristor is very similar to the SUS, discussed above. As the "bilateral" part of the device name suggests, the SBS, unlike the SUS, can pass current in both directions. In this respect, it is quite similar functionally to the somewhat more common and simpler diac (see chapter 4). Like the SUS, and unlike the diac, the SBS has three leads, including a gate lead.

The SBS is a bidirectional SUS, just as the triac is a bidirectional SCR. Essentially, an SBS is just a pair of back-to-back SUSs in a single housing. This is indicated by the schematic symbol used to represent this type of thyristor. The schematic symbol for an SBS is shown in Fig. 12-8.

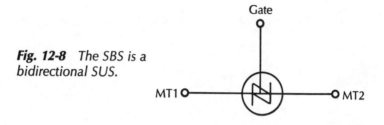

Fig. 12-8 *The SBS is a bidirectional SUS.*

The breakdown voltage for an SBS is relatively low. A typical value is about 8 volts. Like the SUS, the major applications of the SBS are generally in high-speed switching circuits.

THE LASCR

The *LASCR* is an interesting variation on the basic SCR. The name says it all. "LASCR" is an acronym for *Light Activated Silicon Controlled Rectifier*.

Part of the semiconductor is exposed through a clear window in its housing. If enough light is shone on the exposed semiconductor, the LASCR will be triggered. In other words, the light is used as a non-electrical gate connection.

Usually an electrical gate lead is also provided. This lead is normally employed to bias the device, controlling the required turn-ON light level. Figure 12-9 illustrates how a LASCR is usually represented in a schematic diagram.

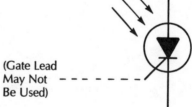

Fig. 12-9 *A LASCR can be turned on by a beam of light.*

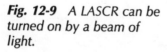

(Gate Lead
May Not – – – – – –
Be Used)

This concludes our examination of thyristors and related devices. You should realize by now that these versatile and powerful semiconductor components aren't so mysterious after all.

Index

Other Bestsellers of Related Interest

BUILD A REMOTE-CONTROLLED ROBOT:
For Under $300—David R. Shircliff

No advanced electronics or computer skills are needed to put together "Questor," a robot butler especially designed by the author to be both affordable and easy-to-build. Shircliff has provided complete step-by-step assembly instructions and photos, wiring diagrams, and parts list. Best of all, when you're through you'll have a robot that looks impressive, *and functions*. 144 pages, 128 illustrations. Book No. 2617, $9.95 paperback, $17.95 hardcover.

OSCILLATORS SIMPLIFIED, with 61 Projects
—Delton T. Horn

Pulling together information previously available only in bits and pieces from a variety of resources, Horn has organized this book according to the active devices around which the circuits are built. You'll find discussion on dedicated oscillator integrated circuits (ICs), digital waveform synthesis, phase locked loop (PLL), and plenty of practical tips on troubleshooting signal generator circuits. 238 pages, 180 illustrations. Book No. 2875, $11.95 paperback, $17.95 hardcover.

500 ELECTRONIC IC CIRCUITS WITH
PRACTICAL APPLICATIONS—James A. Whitson

More than just an electronics book that provides circuit schematics or step-by-step projects, this complete sourcebook provides both practical electronics circuits AND the additional information you need about specific components. You will be able to use this guide to improve your IC circuit-building skills as well as become more familiar with some of the popular ICs. 336 pages, 600 illustrations. Book No. 2920, $19.95 paperback, $29.95 hardcover.

20 INNOVATIVE ELECTRONICS PROJECTS FOR
YOUR HOME—Joseph O'Connell

More than just a collection of 20 projects, this book provides helpful hints and sound advice for the experimenter and home hobbyist. Particular emphasis is placed on unique yet truly useful devices that are justifiably time- and cost-efficient. Projects include a protected outlet box (for your computer system) . . . a variable ac power controller . . . a remote volume control . . . a fluorescent bike light . . . and a pair of active minispeakers with built-in amplifiers. 256 pages, 130 illustrations. Book No. 2947, $13.95 paperback, $21.95 hardcover.

50 POWERFUL PRINTED CIRCUIT BOARD
PROJECTS—Dave Prochnow

If you've ever experienced the frustration and disappointment of failed projects or are interested in finding practical, unique electronic devices, then you won't want to miss this book! Here, in a single reference, are 50 fully described, and detailed electronics projects, complete with schematic diagrams and instructions. More importantly, this book provides what few others on the market can. Each project in this book comes with its own computer generated photo image of the printed circuit board! 208 pages, 184 illustrations. Book No. 2972, $15.95 paperback, $23.95 hardcover.

101 SOLDERLESS BREADBOARDING
PROJECTS—Delton T. Horn

Would you like to build your own electronic circuits but can't find projects that allow for creative experimentation? Want to do more than just duplicate someone else's ideas? In anticipation of your needs, Delton T. Horn has put together the ideal project *ideas* book! It gives you the option of customizing each project. With over 100 circuits and circuit variations, you can design and build practical, useful devices from scratch! 220 pages, 273 illustrations. Book No. 2985, $15.95 paperback, $24.95 hardcover.

TIPS AND TECHNIQUES FOR ELECTRONICS
EXPERIMENTERS—2nd Edition
—Don Tuite and Delton T. Horn

Packed with practical circuit-building tips and techniques, this completely revised and updated edition of a classic experimenter's guide also provides you with ten complete projects. These include a random number generator, an electronic organ, and a deluxe logic probe. Covering such basics as soldering and mounting components, finding and correcting malfunctions and making practical component substitutions, this book also covers the use of breadboards and techniques for finishing your projects. 160 pages, 83 illustrations. Book No. 3145, $12.95 paperback, $19.95 hardcover.

THE LASER COOKBOOK: 88 Practical Projects
—Gordon McComb

The laser is one of the most important inventions to come along this half of the 20th Century. This book provides 88 laser-based projects that are geared toward the garage-shop tinkerer on a limited budget. The projects vary from experimenting with laser optics and constructing a laser optical bench to using lasers for light shows, gunnery practice, even beginning and advanced holography. 400 pages, 356 illustrations. Book No. 3090, $18.95 paperback, $25.95 hardcover.

BASIC ELECTRONICS THEORY—3rd Edition
—Delton T. Horn

"All the information needed for a basic understanding of almost any electronic device or circuit . . ." was how *Radio-Electronics* magazine described the previous edition of this now-classic sourcebook. This completely updated and expanded 3rd edition provides a resource tool that belongs in a prominent place on every electronics bookshelf. Packed with illustrations, schematics, projects, and experiments, it's a book you won't want to miss! 544 pages, 125 illustrations. Book No. 3195, $21.95 paperback, $28.95 hardcover.

Prices Subject to Change Without Notice.

Look for These and Other TAB Books at Your Local Bookstore

To Order Call Toll Free 1-800-822-8158
(in PA, AK, and Canada call 717-794-2191)

or write to TAB BOOKS, Blue Ridge Summit, PA 17294-0840.

Title	Product No.	Quantity	Price

☐ Check or money order made payable to TAB BOOKS

Charge my ☐ VISA ☐ MasterCard ☐ American Express

Acct. No. _____ Exp. _____

Signature: _____

Name: _____

Address: _____

City: _____

State: _____ Zip: _____

Subtotal $ _____

Postage and Handling
($3.00 in U.S., $5.00 outside U.S.) $ _____

Add applicable state
and local sales tax $ _____

TOTAL $ _____

TAB BOOKS catalog free with purchase; otherwise send $1.00 in check or money order and receive $1.00 credit on your next purchase.

Orders outside U.S. must pay with international money order in U.S. dollars.

TAB Guarantee: If for any reason you are not satisfied with the book(s) you order, simply return it (them) within 15 days and receive a full refund.   **BC**